Winemaking

Winemaking

RECIPES, EQUIPMENT, AND TECHNIQUES FOR MAKING WINE AT HOME

Stanley F. Anderson
& Dorothy Anderson

A HARVEST ORIGINAL
HARCOURT BRACE & COMPANY
San Diego New York London

ACKNOWLEDGMENTS

We would like to thank our friend and one of Canada's leading enologists Dr. Elias Phiniotis for reading the technical section of this book and making valuable suggestions. Also, we are grateful for the generous cooperation of the staff of the Federal Department of Agriculture station at Summerland, British Columbia.

Any errors or omissions that remain in the book are the responsibility of the authors.

Library of Congress Cataloging-in-Publication Data
Anderson, Stanley F.
Winemaking: recipes, equipment, and techniques for making wine at home.
"A Harvest book."
1. Wine and wine making — Amateurs' manuals.
I. Anderson, Dorothy, 1929– . II. Title.
TP548.2.A52 1989 641.2′2 88-33876
ISBN 0-15-697095-3 (pbk.)

Original graphics: Sharon Hatelt, Alfred Penz, Colleen Wood
Photography: Dorothy Anderson, Roy Legate
Editing: Giles Townsend
Word processing of manuscript: Lise Dorion
Design: G. B. D. Smith

Printed in the United States of America

K J I

We dedicate this book
to our grandchildren,
GLENNON, FRED, BROOKE, TIFFANY, and JESSICA,
in the hope that they will carry on
the family tradition, encouraged by
occasional sips of Grandpa's wine.

CONTENTS

PART FOUR

PREFACE

Adversity, treated as a challenge, can produce excellence. I can attest to that after forty years of winemaking, a successful business in a field I love, financial independence, and the bonus of a great wine cellar. The secret: No matter how outrageous an idea seems (or how deep the scorn from dear friends), persevere.

My mother has to take some of the responsibility. She made terrible wine. My father, who was a chef on the trans-atlantic steamers, would not touch it, although he would drink the beer she made. Their expertise in the kitchen brought to our table good, wholesome food: Cornish pasties, clotted cream my mother made for the strawberries she grew in the garden, fresh oysters, home-grown pheasant. And the bread! — I can still remember the fragrance. So what was wrong with the wine? Actually, it was pretty typical of what her peers were making in 1935: concoctions from parsnips, dandelions, and potatoes (her potato champagne later became known as jungle juice, with good reason — enough of it turned one into a savage). Still, I loved the occasional taste I was allowed. I was six years old when I became a hero by stealing a bottle for the club potato roast we held in the Wild West of a neighbor's vacant lot. We loved it because it was sweet and it only made you sick if you drank two glasses.

After World War II, when I had become accustomed to the occasional glass of wine with meals, I discovered that my sister had revived the family tradition of making jungle juice. She doesn't consume alcohol, but thought it would help the family budget if her husband and his friends stayed home and drank her homemade wine. I think he gave up drinking for a while. However, she sent me the recipe; so I thought I would give it a try. After three months, the stuff was ghastly — cloudy, with a great deal of sediment. I gave most of it away. A year later, in desperation for something to serve a visiting Latin American consul, I poured him a glass and he pronounced it excellent sherry. I was heartened by this practiced drinker's approval.

I decided to persevere. After all, homemade should be better, shouldn't it? I set myself the task of finding out why wine made at home was so inferior. But I had no idea of the magnitude of the challenge. I first approached some local food technologists I worked with at the time, but they hadn't a clue about winemaking. Then I went to a local winery, where I was rebuffed with Teutonic contempt: "It's far too complicated to make wine at home."

A sensible man would probably have quit right there; for hadn't I been given a professional opinion? But I had doubts. Didn't the word *château* mean "home" in French? And aren't the best wines in the world made in *châteaux* in France?

The first big breakthrough was discovering a book written in England by H. E. Bravery, *Home Winemaking Without Failure.* His technology was not great, but he was an excellent writer and his book introduced thousands to home winemaking in England. Next, I found my mentor, Mrs. Suzanne Tritton, a true professional with a list of degrees after her name as long as your arm. She was British, but trained in the famous School of Enology in Geisenheim, Germany. She not only wrote books for amateurs, but she was a consultant to Harvey's of Bristol. She also had her own laboratory and produced pure yeast cultures for wineries and winemakers all over the world.

Bacchus showered his blessings on me the day Mrs. T. took me under her wing; I went to England as a guest in her home, and we discussed wine from breakfast to midnight, seven days a week.

I returned to Canada to open a shop and act as her representative in North America, distributing her Grey Owl products. Little did I know where it would lead me. In 1961 Mrs. Tritton came to Canada to discuss new products and the Americanization of labels and packaging. As I investigated the North American market, I began to discover significant differences. For instance, Americans and Canadians do things on a broader scale; an Englishman makes one gallon of wine at a time, while an American or Canadian makes anywhere from five to fifty gallons at one time. Early in my career, I realized that for the North American winemaker there are only two quantities of wine: none and not enough!

Another surprise I encountered was that Canadian and American winemakers who start out using rhubarb or plums from their backyard quickly move up to the real thing, vinifera grapes or grape concentrate. A taste for quality is easily acquired.

Here, I felt, was a market to be explored.

When I finally decided to put my dream to the test in the real world of commerce, that's when the meaning of adversity struck home. Sometimes, in retrospect, I think the first indication that you're onto a great idea is when not a soul believes in you. I recall one of my best friends saying to her husband, "Isn't it sad that Andy's giving up his career for that silly idea."

Bankers shunned me like an ex-con with a history of embezzling; my lawyer insisted I was wasting money by incorporating; and to top it all off, the police wanted to shut me down for breaking the liquor laws. Yet five years later, by franchising my know-how, I was able to open sixty stores in ten years. This created over 10 million dollars in sales, thus providing the volume necessary to persuade suppliers to develop new products for our unique market.

For example, I could go to a laboratory that sold 500 yeast cultures a year and offer it orders for 100,000 cultures a year if it could improve shelf life and maintain purity. I went to a scientific-instrument manufacturer who sold 1,000 hydrometers a year and gave him a firm order for 25,000 if he guaranteed precision and a reasonable price.

Perhaps my greatest accomplishment was convincing grape growers in Spain, France, Italy, Australia, and California to put some of their best grapes into grape concentrate instead of just surplus table grapes and winery rejects.

But the most memorable personal triumph came in 1968, when I was invited to lunch in Sausalito by Leon Adams, dean of the American wine scene, author, critic, connoisseur, and the leading authority on striped bass fishing in the Bay of San Francisco. His books on American wines are in the libraries of wine lovers up and down the country. I was unaware of Leon's stature in the industry when we first met, so I brashly offered to supply the wines for our lunch from my own cellar. We started with a sherry-like aperitif, which I had produced accidentally. I was making a full-bodied oloroso, but it turned out too dark and somewhat harsh, so I added charcoal and eventually fined and filtered it. The result was better than I had hoped for, a wine almost identical to a little-known French aperitif, Château Jura, the only wine made in France that develops a sherry flor. Leon correctly identified the wine, which amazed me; but it was his turn to be amazed when he asked me what grapes I had used to produce the wine. I replied, grape concentrate, figs, and bananas. If I recall correctly, he paled somewhat, but bravely carried on to consume my '63 Cabernet Sauvignon, which we shared over the meal.

I relate the story to show that ingredients other than vinifera grapes can on occasion produce an excellent wine. But prejudice is strong, and I am sure if I had told Leon Adams what was in the wine before he tasted it, his pleasure in the wine would have been significantly lessened.

I also know that the wines made by the cellar master who told me in 1958 that it was impossible to make wine at home would have a tough time competing with the quality wines being made today in the homes of Canadians and Americans.

STANLEY F. (ANDY) ANDERSON

INTRODUCTION

This book is dedicated to the lovers of wine, the doers, the scholars, the gourmet cooks, and those who say to us, "I've been thinking about making my own, but it seems so complicated."

Our aim is to remove some of the mystery of winemaking, to make the procedure as clear and simple as possible, and offer you as much information as you choose to read.

Therefore we have divided this book into four sections: "The Basics," for those who want to know what it is all about; "Recipes," for the doers who just want to make a good wine without being burdened by explanations; "Techniques," for winemakers who want a more complete understanding of the equipment and procedures; and "Reference," for those who want to know the whys and wherefores.

The world of wine provides many satisfactions: knowing the varietal grape names, their color and characteristics; being able to correctly evaluate a wine and take a knowledgeable approach to tastings at wineries; knowing which wine to choose in a restaurant and which wine to buy as an investment; and making the great wine districts of the world an important part of your travel plans. But most valuable of all will be the satisfaction of discussing with other oenophiles the production of wine, the grape harvest, vintages, balance, clarity, bouquet, and palate — in short, the inclusion in a very exclusive society.

STANLEY F. (ANDY) ANDERSON
DOROTHY ANDERSON

PART ONE

The Basics

Basic Principles

The basic principles of winemaking have been known and used by man since prehistory. Deprived of oxygen, yeast, a unicellular microorganism, will convert sugar and water to alcohol and carbon dioxide. This reaction — the way in which yeast obtains energy for growth and reproduction — is the basis of all winemaking. Much of our task as winemakers, therefore, consists of making sure the yeast has what it needs to get the job done.

Yeast grows best in a solution of 22% sugar; so this is the environment the winemaker tries to provide, either in the naturally occurring sugar of a fruit pulp, or in fruit pulp made up with additional sugar to a concentration suited to the yeast's liking. Yeast also needs minerals, nitrogenous compounds, and B vitamins; consequently the winemaker makes sure these are present, either in the wine's basic fruit ingredient or as a specially prepared additive, a yeast nutrient. Similarly, because yeast grows best between 60° and 80°F (15–26°C), this is the temperature range at which the winemaker tries to keep it, at least until most of its work is done.

But wine yeast must not only be fed, cossetted, and nurtured; it must also be protected from competition. There are hundreds of strains of yeast, most of them wild, and only some of them capable of producing alcohol without unpleasant tastes and by-products. The air is filled with these cells, and until the nineteenth century winemakers relied on such airborne invaders to get their wine going. As a result, the practice of winemaking was never entirely certain; a wine might not ferment at all, or it might be colonized by a strain of bacteria that turned the whole batch to vinegar. Nowadays, winemakers leave nothing to chance. Strains of yeast are systematically bred and selected, and great care is taken to inoculate wine with large numbers of cells of the right kind with the right properties. Nevertheless, no matter how well established a yeast might be, a wine is never safe from the countless airborne bacteria and mold spores only too eager to colonize the rich nutrient environment it affords. For this reason, winemaking containers and utensils must always be cleaned and sanitized before use, and every container of wine must be kept covered or sealed with a fermentation lock.

For all its efforts, the task of the wine yeast is ultimately thankless. Yeast produces alcohol, but it cannot live in alcohol — at least, not beyond a certain point. Thus, when a concentration of 10–19% alcohol is reached, the yeast can no longer tolerate it; its activity comes to a halt and fermentation stops. The winemaker then filters the wine to remove any sediment and places it first in a bulk container, then in bottles, for its proper period of aging.

We have talked about yeast, but fermentation is only part of the winemaking process. Under the right conditions, any fruit can be fermented, but the result is not necessarily wine. Its fitness for that description is governed by certain criteria of taste, aroma, and physical appearance that the winemaker spends a great deal of ingenuity striving to attain. By wine, we mean a fermented beverage of 10–14% alcohol, pleasing to the palate, clear in appearance, and with a perceptible, but not obtrusive, taste and aroma of its original fruit ingredient. But there are factors in many basic wine ingredients that work actively to prevent this ideal end product. These factors must be anticipated, detected, and neutralized by the winemaker before a good wine can be made.

For instance, many fruits have a high acid content; high acid produces wine that is sour and harsh. Too little acid, on the other hand, produces wine that is insipid and flat. The careful winemaker measures the acid content of his ingredients and reduces the acid or makes up its deficiency as needed. Many fruits contain high quantities of pectin, a substance that inhibits flavor and leads to cloudiness or jelly-like lumps in wine. When using high-pectin fruits, therefore, it is customary to add an enzyme that breaks down pectin. Tannin too is often added, to provide body and astringency and improve the clarity of a wine. Other substances, such as glycerine and oak chips, are added purely to supply a pleasing consistency or interesting undertones to a wine's bouquet.

A wine also undergoes many spontaneous and subtle changes after fermentation is complete. This is why wines are allowed to age. Calculating the exact time a wine should be left to sit in a cask or bottle is a matter of the utmost care and deliberation, and takes account of the basic ingredients and the type of wine the recipe is intended to be.

We can classify wines in several ways. We can group them by color, into reds, whites, and rosés — the color depends on the color of the basic ingredient fruit but also on the process used. We can group them by sugar content — a dry wine has less than 1% sugar; a sweet wine has 1% or more residual sugar and has a detectable sweetness. And we can classify them according to use.

An *aperitif* is slightly sweet, high in alcohol, and drunk as an appetizer before dinner. Sherry is an aperitif. *Table wine* is intended to be drunk with a meal. Because its taste must complement the taste of the food, it is a dry wine, containing 9–12% alcohol by volume. Generally, strong-tasting food, such as red meat or spicy cuisine, requires a red wine; food with a more delicate flavor, such as fish, requires a white wine. Rosé will usually go with either type. *Dessert wines*, as their name implies, are wines that go with the last course of the meal. They are sweet and designed to complement fruit or sweetened dessert.

Enologists (wine scientists) also classify wines as *reductive* or *oxidative*, a distinction that refers to the role of oxygen in their aging process. Reductive wines are permitted little or no oxygen and are generally drunk young; oxidative wines arc aged so as to expose them to minute quantities of oxygen over a long period of time.

Each wine requires a treatment that will bring out the characteristics suited to its ingredient and purpose and eliminate those that are not. In this way winemaking is like a careful juggling act. But it need not be difficult. Our recipes are tried and tested. Follow them closely and observe a few simple precautions, and the principles of winemaking are at your command.

Basic Equipment

Do not use make-do equipment for winemaking. In North America and Western Europe, equipment designed specifically for winemakers is readily available from specialty winemaking stores. It is both reliable and cheap. Our emphasis will be on 5-gallon U.S. (19 lit), or 25-bottle, batches of wine. The proper equipment to make this amount will cost you between $50 and $60 — no more than a dozen good wineglasses or a case of drinkable wine.

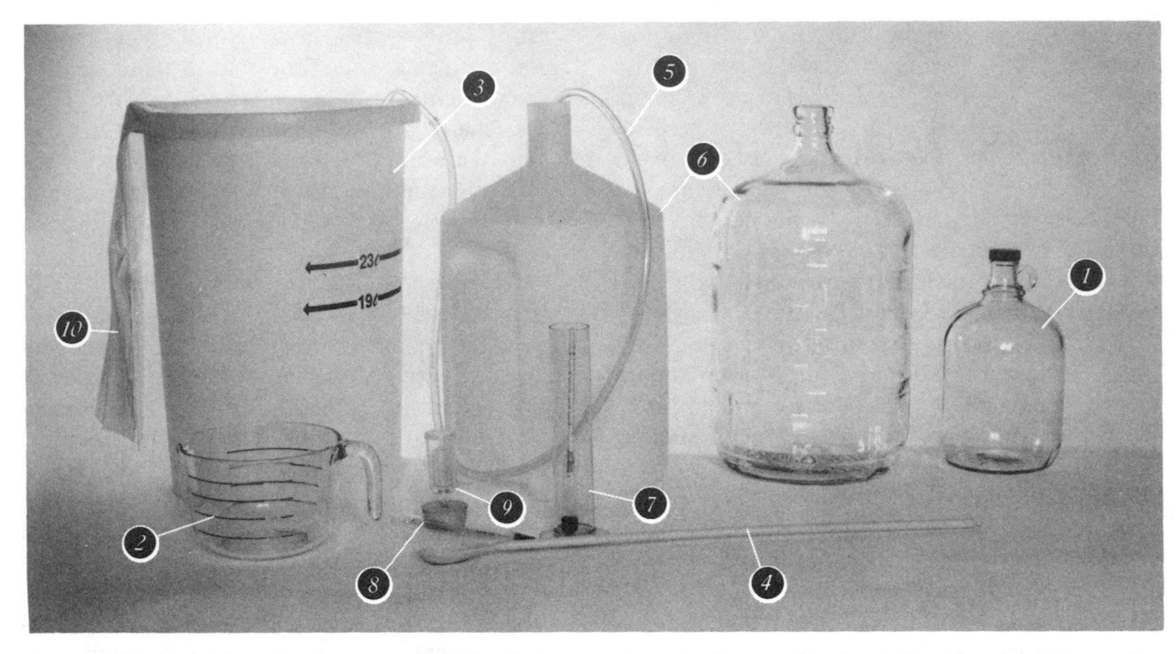

The ten basic items of equipment for the home winemaker: 1 — glass jug with a screw cap, 2 — pitcher or pail with graduated measurements, 3 — large food-grade plastic tub, 4 — long-handled wooden or plastic spoon, 5 — syphon hose with plastic tip, 6 — two glass or plastic carboys (one of each shown here), 7 — hydrometer set, 8 — floating thermometer, 9 — two fermentation locks with bungs (one shown here), 10 — plastic sheet

We begin with the Basic 10 — ten basic items of equipment that your small winery will need. If well looked after, they will give you many years of use.

THE BASIC 10

1. Glass jug with a screw cap, to hold sanitizing solution
2. Pitcher or pail with graduated measurements, for the accurate measurement of water
3. Large food-grade plastic tub, to be used as a primary fermentor
4. Long-handled wooden or plastic spoon
5. Syphon hose with a plastic tip
6. Two glass or plastic carboys
7. Hydrometer set
8. Floating thermometer
9. Two fermentation locks with bungs, to fit the carboys
10. A plastic sheet

1. **Glass jug** (1 gallon) for sanitizing solution. It is essential to clean all equipment with a sanitizing solution prior to each use. This solution rids equipment of any wild yeasts or bacteria that could spoil your wine. The solution will keep well in a glass container, provided it remains tightly capped.

2. **Pitcher or pail** (2 liters) with a graduated scale. This allows the precise measurement

of water, sugar, and so on. Winemaking is an exact science; measurements must be accurate.

3. **Large food-grade plastic tub**, to be used as a primary fermentor. As the name implies, this is the container where the first part of the fermentation takes place. Why food grade? Most plastic pails and tubs are colored to make them more attractive. A study conducted in England showed that it is possible for a mixture of alcohol and acids to leach out heavy metals contained in the dyes. Food-grade plastic is white, translucent, or transparent. Even food-grade plastic sometimes gives an off odor when new. We recommend that you fill the container with warm water and a tablespoon of baking soda and let it soak overnight. Rinse with clear water before use.

4. **Long-handled wooden or plastic spoon** for stirring ingredients. Do not use metals of any kind in winemaking, except stainless steel. The long handle of the spoon is useful for stirring additives into narrow-necked carboys.

5. **Syphon hose**, to transfer liquids from one container to another. You need 5 feet (154 cm) of 3⁄16th (8 mm) inner-diameter clear plastic hose, attached to a 22-inch (56 cm) length of rigid plastic tube. Syphoning shields the wine from contact with the air and allows you to empty your fermentor without disturbing the sediment.

6. **Two glass or plastic carboys**, to be used as secondary fermentors. The word *carboy* comes from the Persian word *qarabah*, meaning "a large glass container to hold corrosive liquids." Wooden barrels can be used as secondary fermentors, but for 5-gallon (19 lit) batches, glass or plastic is best. One-gallon glass jugs are serviceable, too, but the 5-gallon carboys of glass or plastic are less work.

For larger quantities there are demijohns from Europe that hold anywhere from 25 to 64 liters. These attractive containers used to come in wicker or raffia baskets, but now, due to import restrictions, they come padded with plastic. When available, they cost $35 and up.

The key feature of the secondary fermentor is that it enables you to exclude air from the gently fermenting new wine. You cannot use a 14-gallon carboy as a secondary fermentor for 5 gallons of wine — the airspace is too great and your wine will spoil. You must be able to fill your secondary fermentor to the top, leaving only a small surface area at the neck of the bottle, in which you must insert a fermentation lock. Ideally, a secondary fermentor should be filled to within 2 inches (5 cm) of the fermentation lock and maintained at that level. We will be repeating the warning "Top up" at every opportunity throughout the book. Remember that the quantity of wine you make must be tailored to the size of your fermentors — both primary and secondary.

For red wines, ports, and sherry wines, which we classify as oxidative, either plastic or glass secondary fermentors are acceptable. For whites and rosés, which are classified as reductive, we prefer to use glass fermentors only. Food-grade plastic is porous; like wood, it allows the wine to breathe minute amounts of oxygen, which helps red wine to mature. But this also means that you should not leave your red wine in plastic fermentors too long. Start to smell and taste it after 3 months; it will normally take 6–9 months to age, but if your carboy is thin or you store it in a warm place over 70°F (21°C), the process is faster — not better, but faster.

If you wish to use wooden barrels for your secondary fermentation, acquaint yourself thoroughly with the procedure for barrel use described on page 244. Great care is needed to use barrels successfully.

Note: To speed the emptying of carboys when rinsing, turn them upside down and swirl their contents in a rotary motion. The liquid will empty out in a smooth stream.

7. **Hydrometer set.** You will need a hydrometer and a plastic testing jar to measure the density of your must and your wine. This simple instrument will tell you if you have too much or too little sugar to start. That in turn lets you control the amount of alcohol the fermentation will produce. A hydrometer puts you in charge of the single most important ingredient in your must — sugar.

8. **Floating thermometer.** This can be left in the must to give you a constant temperature reading. Too high a temperature will kill the yeast; too low a temperature will inhibit fermentation.

9. **Fermentation locks**, with bungs to fit your carboys. Fermentation locks, also known as bubblers or breathers, are low-pressure valves that allow carbon dioxide to escape, but prevent air from getting into your wine. Bungs are rubber or polypropylene stoppers that wedge tightly into the neck of the carboys. Try to purchase the white bungs because black rubber sometimes leaves a rubbery smell in your wine. Cork can also be used to hold fermentation locks in place, but cork deteriorates and does not give you as airtight a closure.

When you insert a bung, use a paper towel to make sure both it and the inside of the carboy neck are dry — otherwise the bung may pop out.

10. **Plastic sheet.** This is used to cover the primary fermentor. Tied down, it keeps airborne bacteria and fruit flies out of your must. Do not use large garbage bags to cover your fermentor. Many now are treated with an anti-odor or animal-repellent chemical — this could be a toxic hazard.

THE WELL-EQUIPPED WINERY

As you gain experience in winemaking, you will want to undertake more ambitious recipes. This list completes the equipment you will need for most winemaking projects.

1. Acid-testing kit
2. Gram scale
3. Funnels
4. Straining bags
5. Filter and pads
6. Hand or bench corking machine
7. Bottle washer
8. Bottle draining stand
9. Corks, labels, foils, and capsules
10. Sugar refractometer
11. Alcohol refractometer
12. Wine thief
13. Grape crusher and destemmer
14. Grape press
15. Electric pump and hoses
16. Oak barrels
17. Barrel stands
18. Heating pad or belt
19. Sulphiter
20. Sulfikit™

1. **Acid-testing kit.** A testing kit allows you to measure the acid content of your must or wine in grams per liter. It contains 2 syringes graduated in milliliters, a 50-millimeter vial, sodium hydroxide, phenolphthalein, and litmus paper.

2. **Gram scale.** A scale is very helpful when measuring small amounts of dry additives. It saves buying expensive tablets.

3. **Funnel.** A large funnel is needed to fill carboys and barrels; a small funnel is needed to fill bottles.

4. **Nylon straining bag.** A straining bag allows you to remove pulp from fruit wines and

Some of the supplementary items of equipment for the home winemaker: 1 — gram scale, 2 — filter and pads, 3 — sulphiter/bottle draining stand, 4 — grape crusher and destemmer, 5 — grape press, 6 — electric pump and hoses, 7 — oak barrel (here of 40-gallon capacity)

coarse-strain the "free run" in the fresh grape process.

5. **Filter**. A pressure filter using a commercial-quality filter pad or membrane allows you to remove particles suspended in your wine. Several brands are available; Vinamat and Polyrad are well known and effective.

6. **Corking machine**. A corking machine is essential to insert real wine corks. The model you choose will depend on the volume of your annual production. A hand corker is sufficient for 100 bottles. For 300 bottles or more, a bench or floor model is a necessity.

7. **Bottle washer**. This device threads onto the tap. It saves hot water and time, and washes everything, from pints to carboys.

8. **Bottle draining stand**. A draining stand is a good idea if you are short of space.

9. **Corks, labels, foils, and capsules**. Corks protect bottled wines; labels, foils and capsules dress up wines for the table or for giving as gifts.

10. **Sugar refractometer.** This clever device is for the vineyardist. With one drop of

juice it allows you to calculate the sugar content of your grapes.

11. **Alcohol refractometer**. With an accurate hydrometer and an alcohol refractometer, you can measure the alcohol by volume in your wine very accurately. Not for everyone — expect to pay over $200.

12. **Wine thief.** A glass or plastic tube that allows you to withdraw small quantities of wine through the narrow opening of a carboy or barrel.

13. **Grape crusher.** We recommend a hand-driven crusher with the capacity to handle one lug (35 lb/16 kg) of grapes at a time. This will allow you to process 500 pounds (227 kg) quite comfortably. A grape crusher is also good for crushing cherries, but it will not crush apples; for apples you need an apple crusher.

14. **Grape press**. We recommend a basket press with a threaded steel post. Presses are rated by the pound or kilo of crushed grapes they can press at one time. You need one that handles at least 200 pounds (90 kg) at a pressing. The Italian imports are numbered 25 kg, 35 kg, 45 kg, and so on; do not buy one smaller than 35 kg. If you think you need a smaller one, you don't really need a press. A grape press does not work well for apples; for apples you need an apple press.

15. **Electric pump**. If you make batches of 10 gallons (40 lit) or more, you should consider a plastic pump and 30 feet (9 m) of food-grade plastic hose. You can buy them separately and assemble them yourself for around $100. A pump saves time and backaches.

16. **Oak barrels**. Barrels require care, but they make great red wines.

17. **Barrel stands**. If you use barrels, you will need barrel stands. Barrels of wine must lie on their sides, preferably 2–3 feet off the floor.

18. **Heating pad or belt.** Heating pads may be purchased from drugstores; heating belts are available from hardware stores and wine shops. The belts can be strapped on the sides of fermentors, and the pads can be placed under them. The belts are designed to keep motors and pipes from freezing. They prevent stuck ferments when fermentors are stored in cool areas.

19. **Sulphiter**. This device squirts sulphite solution into bottles prior to bottling — a time-saving gadget.

20. **Sulfikit™**. This is used to measure free sulphur dioxide in wine or juice.

Basic Ingredients

By basic ingredient, we mean the principal fruit that will dominate the flavor and aroma of a wine. Let's start by postulating that the word *wine* means the fermented juice of the vinifera grape. All other fermented products should have a different name or a modifying adjective, such as rice wine or sake, blackberry wine or concord grape wine. This is not to say that wines made from berries or tree fruits are not good — a well-made blackberry wine will be as good as or better than an ordinary wine from vinifera grapes. But there are important differences.

You can crush grapes, place them in a closed container, and nature will make wine; not great wine, but drinkable wine. If you crush blackberries and give them the same treatment, there is a good chance they will not ferment at all. And if they do ferment, you'll be lucky to get 6% alcohol in a sour, cloudy liquid disfigured by blobs of pectin jelly — a potion that would not even make good wine vinegar. To understand what is happening here, we must look at the chemical composition of the two fruits. Blackberries contain 10% sugar and 6–7 aromatic constituents; they are also high in pectin and low in tannin. Ripe grapes, on the other hand, contain 20–24% sugar and 22–24 aromatic constituents; they are low in pectin and high in tannin. This distinction is critical. Yeast likes a fairly high concentration of sugar, around 22%; low aromatics means a dull flavor, while high aromatics means a complex and satisfying flavor; furthermore, pectin forms lumps and inhibits flavor, while tannin is a natural clearing agent for wine and provides a pleasing body and astringency.

Thus, when we reserve the word *wine* for the fermented juice of the vinifera grape, it is no historical accident. The juice of the grape provides exactly the conditions in which yeast can produce a high concentration of alcohol and leave behind a smooth, clear, fine-tasting beverage. To that extent, it serves as a model for all our recipes using other fruits. When we write a recipe using a basic ingredient other than wine grapes, our objective is to produce a must (the crushed fruit or juice prior to fermentation) with a balance of constituents as near to the wine grape's as possible. To copy the grape's acidity, we add tartaric acid. To reduce an excess of other acids, we add water. To ensure adequate alcohol, we add sugar until it reaches a grapelike 20–24%. To make good any deficiency of minerals and vitamins that the wine grape provides, we add yeast nutrients. Following the same principle, we add tannin; and when we use fruits high in pectin, we add pectinase, an enzyme that breaks down its long, binding molecules.

But the fact that we can simulate the grape's desirable fermentation properties with additives does not mean we should be indiscriminate. Many authors suggest that you can make wine from anything, including onions, pea pods, old jam, and grass clippings. Their maxim is "If you have a surplus of anything and can't think what to do with it, turn it into wine." We cannot subscribe to this view. Old jam and preserves are very likely to be oxidized; and when food or wine is oxidized, it's spoiled. Onions, pea pods, and so on have little or no fermentable sugar or fruitlike flavor; thus, they cannot be considered a basic ingredient. Our advice is, put them where they belong — into the soup pot or the garbage can.

We have divided our recipes into groups of ingredients that can be treated similarly: berries, soft fruits, hard fruits, tropical fruits, dried fruits, grape concentrates, and fresh grapes. We follow with sections on champagne and liqueurs.

BERRIES

Blackberries, saskatoon berries, elderberries, loganberries, blackcurrants, blueberries, cranberries, raspberries, and strawberries all make good red wines.

Blackberry wine, perhaps the best of the berry wines, resembles a Bordeaux if properly aged, and makes a pleasant social wine if slightly sweetened. It can be consumed after 6 months. Saskatoon berries, native to the western prairies, are similar in flavor and make a good dry red table wine.

Elderberries make an aromatic red wine. However, those grown in the coastal areas of North America are generally to be avoided. They tend to be bright orange and contain a bitter oil that will float to the surface of your fermentor and impart an acrid taste. The bluish-black elderberries that grow inland are the ones to use. Dried elderberries, available in specialty winemaking stores, make an excellent additive to red wines made from grape concentrate because they supply the tannin concentrate lacks.

Loganberries and blackcurrants make delicious sweet dessert wines. Blackcurrant wine has an aroma similar to Cabernet Sauvignon.

Blueberries grown on the East Coast make an excellent dry red one surprisingly similar to a European table wine. Blueberries grown on the West Coast, however, contain a natural source of potassium sorbate, which inhibits fermentation. If you have access to blueberries only of West Coast origin, add grape concentrate.

Wines made from cranberries, raspberries, and strawberries have a delicate flavor and, when sweetened, make a refreshing drink over ice on a summer's afternoon. Dry wines from these berries are less successful.

Recipes using berries begin on page 31.

HARD FRUITS

Apples, pears, and rhubarb, which are classified as hard fruits, can be made into a light, refreshing wine without the use of a crusher or press. Bartletts are our favorite pear; and Gravensteins, blended with Delicious, are what we usually use for our apple wine. We love cider, but unfortunately our apples are usually ready when our fresh grapes are about to arrive, and the apples take second place. If you have access to an apple crusher and press, we include a recipe for champagne cider.

Recipes using hard fruits begin on page 56.

SOFT FRUITS

Apricots are a versatile fruit for winemakers. They are bland enough to make a delicate table wine that often passes as a grape wine, yet they make rich, full-bodied liqueurs and a pleasant champagne. They ferment easily because they are rich in the organic nutrients required by wine yeast; consequently they serve as a good base for fortified wines and make an excellent vermouth. If no wine grapes were available, and we were obliged to pick one fruit for white wines, we would choose apricots.

Because plums grow in abundance all over North America, many novice winemakers use them for their first batch of wine and are sadly disappointed. This is because red and blue plums acquire an odd, spicy taste during fermentation that is not pleasant in a red table wine. White and yellow plums do not have so strong a flavor; nevertheless, we suggest you use your plums for sherries and ports only. Like apricots, plums ferment well and, with high-alcohol yeast strains, make a good base for fortified wines. Among the thousands of wines brought to us by winemakers for evaluation, two were truly outstanding: a plum port that had been aged 5 years, and a plum Madeira baked and aged 3 years that tasted like a full-bodied sherry.

Peaches make a delicate white wine, but they are deficient in yeast nutrients; consequently they do not ferment well. Be sure to use the grape concentrate called for in the peach wine recipes.

All types of cherries ferment well, but we prefer Bings. The white Saint Anne's and the sour morello cherries are less successful because, like plums, they tend to develop an off flavor during fermentation. The sweet red varieties can be used for red table wine; slightly sweet social wine; quinine-flavored wine, like Saint Raphael; and full-bodied port. Like plums and apricots, cherries contain the right nutrients to ferment to high alcohol levels. A case in point: A gentleman brought us a bottle of his

2-year-old cherry wine for evaluation. He was worried; to him it tasted terrible, and he had two 40-gallon barrels of it. We found it was clean on the nose, but very harsh on the palate. We measured the alcohol content and checked it twice before arriving at a reading of 19.8% alcohol by volume — half as strong as whiskey. We sweetened it to 6% residual sugar and found we had an instant delicious ruby port. The gentleman now had a lifetime's supply of superb port wine, 400 bottles.

Recipes using soft fruits begin on page 72.

TROPICAL FRUITS

Kiwi fruit and mangoes, once displayed as expensive and exotic treats, are now commonplace in many parts of the country and are often advertised as specials. Although these fruits are not cheap, our recipes need no more than 10 pounds or so and make very satisfying German-style wines. Bananas, pineapples, lemons, oranges, and grapefruits are everyday fruits in our supermarkets, but do not make good table wines, being better as sweet social wines. Prickly pears also make a good social wine.

Recipes using tropical fruits begin on page 101.

DRIED FRUITS

Dried fruits are available in supermarkets worldwide. Where there are no fresh fruits in abundance, we recommend dried fruits such as raisins, figs, and dates for high-alcohol dessert and social wines. But apart from dried vinifera grapes, which are available only at winemaking outlets, dried fruits do not make good table wines.

Recipes using dried fruits begin on page 121.

VITIS VINIFERA

Wine from grapes has been part of our diet for several thousand years. It is thought viticulture was practised by the Egyptians and Assyrians as early as 3500 B.C. and by the Greeks in 1400 B.C. The popularity of wine spread rapidly, and by the fifteenth century grapes were widely grown throughout Europe. In the eighteenth century, European explorers brought cultivation techniques to the New World.

The most important wine-producing area of the world is still Europe, which has more than 70% of the acreage, but wines from Australia, Chile, South Africa, and the state of California are becoming increasingly well known. For the North American home winemaker, California grapes are the obvious choice.

The most popular California wine grapes (by wine sales) are as follows:

Cabernet Sauvignon. These small tough-skinned grapes are considered the best of the California reds. To reach its peak flavor, their wine should be aged 1–3 years in the cask, and at least 4 years in bottles.

Pinot noir. Perhaps France's finest wine grape, Pinot noir has produced widely varying results since being transplanted to California. It is often blended with Petite Sirah for extra color and tannin, and has a rich, complex taste if allowed to develop its full character in the bottle. This may take from 3–6 years.

Chardonnay. With a flavor reminiscent of apple, this grape is used in the renowned white Burgundies Chablis and Pouilly Fuissé. Chardonnay wines age well and have become very popular.

Chenin blanc. Used as a blend in many wines, Chenin blanc is fruity with a touch of sweetness. It is high in acidity, but its wines do not age well and should be consumed within 1–2 years.

Johannisberg Riesling. Riesling grapes are harvested late and their wine should never be completely dry. Lightly scented with a soft fruit flavor, they are used in many variations of lush-tasting wine.

California Zinfandel grapes also are excellent if grown in the table-wine grape areas — Napa, Sonoma, and Monterey being the most notable. Zinfandel grapes do not restrict you to a red wine; very popular now is white Zinfandel table wine. This is produced by lightly crushing the grapes and pressing them immediately so that the juice remains white; it is then fermented out and treated as white wine. The remaining skins and pulp can then be fermented to make a red. We recommend making both white and red wines from Zinfandel, Cabernet Sauvignon, Merlot, and Petite Sirah. The whites are drinkable within 3–6 months, as opposed to 3–5 years for the big reds.

Unfortunately, California grapes do not travel well, Chardonnay especially. If you live in the North or East and prefer white wines, you might explore the availability of the Pinot Chardonnay grown in New York state, or the Riesling and Gewürztraminer grown in Washington state.

For information on ordering grapes, see "Purchasing Fresh Grapes," page 263.

FRENCH HYBRID GRAPES

Climate is the principal factor controlling vine-growing. Growers in Canada and the eastern United States, which do not have the Mediterranean climate necessary for the cultivation of *Vitis vinifera*, have struggled ingeniously to solve the problem by crossing *Vitis vinifera* with *Vitis labrusca*, a cold-weather vine native to North America. The hardy strains that resulted are commonly called the French hybrids. Originally designated by number because of the multitude of varieties produced, the more successful have now been given exotic names, such as de Chaunac, Chelois, Foch, and Thurgau. However, these hybrids can claim only partial success, since their grapes are high in acid and low in sugar; and unless they are ameliorated by the addition of California grape juice, they produce a wine high in acid, with a sour taste. Painstakingly, the search goes on for the perfect hybrid that will produce the magic numbers of the varietals' chemistry: 22% sugar, 6.5 g/lit acid for the reds; 21% sugar, 7.5 g/lit acid for the whites.

Meanwhile, if you have a choice, use hybrids for white wine only and ameliorate the must either by removing some of the acid or by adding grape concentrate. (See "Hybrid Grape Use," page 262.) If you wish to make red wine from hybrids, then it is even more important to use grape concentrate. Also, be careful of fermenting the grapes with stems in the must. Hybrids are often slightly underripe at harvest, and their stems make a "stemmy"-tasting wine. They are also high in pectins, and a pectic enzyme must be added at the time of crushing.

New varieties are being developed constantly, but the following have been well tested throughout the last 20 years and are accepted by Canadian wineries:

Whites	***Reds***
OK Riesling	Rougon
Müller-Thurgau	Chancellor
Verdelet	

Recipes for fresh grapes begin on page 181.

FRESH GRAPE JUICE

Some alternatives to fresh grapes have recently appeared on the market.

Some years ago, grape importers began to make fresh grape juice available to local buyers during the grape season. They would crush and press the grapes themselves and sell the juice in 5-gallon pails. Some vendors suggested that buyers add wine yeast; others simply advised their customers to let natural yeasts take over. The result was often very poor wine or good vinegar.

One of the drawbacks was that you could make only white wines; to make red wines you have to ferment the juice of red grapes with their skins. Some grapes do have red juice, but they are not quality wine grapes. Heating the crushed grapes was a method sometimes used by importers to extract the color from the skins, and it killed the wild yeast; but it did not extract enough tannin from the skins, stems, and seeds. Tannin is essential to good red wine.

Later, growers began to extract juice at the winery and ship it to buyers under refrigeration. This sounds good until you realize that some yeast cells can multiply at temperatures as low as 28°F, and most of this juice was stored at 35–38°F. Consequently, much of it was sold half fermented. This is not the way to produce good wine, because it does not allow you to test and balance the must before fermentation. Besides, wild yeast seldom gives as good results as selected wine yeast. Nevertheless, if the juice had been stored cold enough and contained 50–100 ppm sulphur dioxide — and if you received it soon after crushing and pressing — this juice could make a good wine. But you still needed to add tannins.

The next product on the market was frozen must, complete with grapes. This was available to the consumer as a 6-gallon block sealed in a box. We were very enthusiastic about this product, but unfortunately, the Cabernet Sauvignon we tested in the lab had a vegetable-like nose. It had come from Monterey in California, and apparently, much of the commercial wine made in that area had the same problem, so it was hardly a fair test. One of the problems with frozen must is the high cost of shipping and storage. And of course it has to be used as soon as it thaws.

In the 1980s, food technologists came up with aseptic packaging — exposing a product to a very high temperature and then sealing it in a sterile container. We have tested aseptic grape juice packed in 5-gallon bags for 3 years. We have tried Cabernet, Zinfandel, Riesling, Pinot, Chardonnay, and French Colombard. So far, the results are very promising. The cost is a trifle high at present; but as sales increase, prices could come down. The U.S. price at the time of writing is $6 per gallon — $9–$9.50 for Canadian winemakers, with freight charges.

While there is still a lack of tannin in the reds, the whites are very good. When you open the package, the aroma of fresh grape juice fills the room, even after being stored 6 months at room temperature. It's so fresh-tasting, it's hard to believe it does not require refrigeration.

To some extent, aseptic-pack fresh grape juice will cut into the market for fresh grapes because it is so much easier to use. You don't need a crusher or a press, and there's no fussing with the disposal of grape skins and stems. You can make your year's requirements of wine over a 3- to 6-month period. It's almost as convenient as grape concentrate. In fact, our favorite way of using aseptic-pack fresh grape juice is blending it 50-50 with grape concentrate. If you use top-grade concentrate, you reduce the price of each bottle of wine without noticeably diminishing the quality.

At the time of writing, these products are available in western Canada and the U.S. They should be available everywhere soon.

CONCENTRATES

Grape concentrate is simply grape juice from which water has been vacuum-removed at body temperature. Only the water is taken out; all the flavor and character of the premium grapes remains. Modern technology has removed the hard work and the mess.

The arrival of Australian grape concentrates on the North American market has been a giant step forward for winemakers accustomed to an industry in which progress has been uneven.

In the late 1950s a company in Spain started packaging high-density grape concentrate in 40- and 80-ounce cans. Shortly thereafter, another Spanish processor started shipping grape concentrate in 40-gallon wooden barrels. Both these products were avidly accepted by amateur winemakers in England and Canada. The ease of their use persuaded consumers to overlook the fact that if they were not used within a month of their arrival from Spain, the wines they made developed a metallic taste.

By 1970 concentrate producers had recognized the potential of sales to amateur winemakers, and they designed products to capture the growing market. The most difficult fact for them to comprehend was the amateur's interest in quality and the demand for varietals. We recall the remarks of a prominent chemist in a California concentrate plant who said, "Home winemakers want an alcoholic beverage they can drink in three weeks; they don't care about quality." He was wrong on both counts. Home winemakers are the largest single body of experts when it comes to judging wine.

With the emergence of varietal concentrates from a plant in California, buyers rapidly became interested in the grape variety of the concentrate they were buying. California wine laws specified that a varietal wine must contain 51% of the varietal name on the label. The California concentrates generously contained up to 70% of the varietal named. For the most part, they were a success; they were fruitier than their European competitors, and had better color. This led to an improvement in the products from Europe and other suppliers. Now, when a producer offers concentrate to importers, he has to verify the variety of its grape. We see Malbec from Argentina, Müller-Thurgau from Austria, Zinfandel from California.

Concentrates from many countries are now available: France, Italy, Portugal, Cyprus, Spain, Argentina, Chile, Brazil, Austria, and Australia. In most cases, the only grapes that go into concentrates are high sugar content grapes that are surplus to winery demands.

Recently, more California growers with modern equipment have become involved in producing high-quality varietal concentrate. The reasons are twofold: Not only have they become aware of the huge market of home winemakers, but an increasing number of small wineries are now basing their production on top-quality grape concentrate.

If you live in the USA, or just love California wines as we do, don't hesitate to try the concentrates your retailer suggests. He will know the varieties used, where the grapes were grown, and when they were harvested. Remember, quality-packed varietal grapes are never cheap. When your supplier offers you a discount, ask whether it's last year's harvest instead of the current vintage concentrate. California vintage is usually available in October, European in December, and Australian in July. Concentrate stored at a cool 38°F (4°C) will keep for up to a year; but concentrate stored at 70°F (19°C) will deteriorate in 3 months. Not all suppliers know or care to admit this fact.

Remember, white concentrate should be a very pale gold; red concentrate should be blue-red, not brown. Where you are sure no artificial color has been added, color is your best indicator of quality in grape concentrate. Never, never, buy brown concentrate for table wine — no matter how cheap it is. At best it may make sherry or Madeira.

No concentrate supplier has ever admitted to adding color to its red concentrate, but among the hundreds of samples we have examined in our lab, there were a few that have caused suspicion — especially when the color would not come off the lab sink!

It is not easy to compare prices of concentrate unless you can obtain unit prices; that is, cost per ounce, plus the density. The international standard for grape concentrate is 68–70 Brix, or 68–70% solids, which are mostly grape sugar. Some concentrates are as low as 30% and others are as high as 77%. Concentration above 70% often damages the flavor and is therefore undesirable.

Finally, we're back to the most recent arrival on the market — what we choose to call the Australian Collection. Australian concentrates are definitely gem quality, and given an honest evaluation, they will change winemakers' attitudes to wines made from grape concentrates.

Harvested and concentrated at their peak of maturity by the most advanced technology available, they bring you a grape juice so natural in color and taste, it's like standing in the vineyard and squeezing fresh grapes in your hands.

The white concentrates are clear white with a touch of greenish gold; the reds are red with a hint of blue — just as fresh juice should be. At last, you can make a Cabernet from concentrate, and everyone will be able to detect the aroma of Cabernet Sauvignon in the glass. And if you are accustomed to apologizing for white wines that are made from concentrate, you're in for a joyous surprise. Now you can make Rhine- and Burgundy-style white wines that look and taste as they should — clear, with a glimmer of green and a ray of sunlight. This appearance, combined with the true varietal flavor and aroma of the genuine Johannisberg Riesling, means you won't have to add the word *-type* when you introduce your Hock.

White wine has become increasingly popular over the past 5 years, even among connoisseurs. But there are times when only red wine is right. Consequently, a new class of red wine has emerged for the white-wine devotee: red rosé and chillable red — popular names for light, fruity red wines that taste good without 5 years' aging and an expensive price tag. The Australian Collection includes two fancy varietals that are perfect for this new category of red wines: Malbec, a grape that immediately brings to mind the currently popular French imports, and Grenache, a grape famous for rosé and red rosé, reminiscent of the popular Verde wine of Portugal.

Concentrates and Blanc de Noir

Blanc de noir is made from red vinifera grapes. It ranges in color from a slight pink blush to rosé. There are four popular varieties on the market now, labeled white Zinfandel, white Merlot (from Italy), white Cabernet, and white Pinot noir. Very few actually carry the label *blanc de noir.*

The success of white wines from red grapes depends on the sugar and acid balance. They must be harvested before the customary date for reds; the acid must be higher, the sugar lower. Also, they must be pressed lightly. After pressing, the pomace (grape pulp) can be used to make an excellent red wine with the addition of grape concentrate. Thus, with the same grapes you can fill your cellar with both white and red wines! With 420 pounds (190 kg) of Zinfandels, you could produce

12 gallons (45 lit) of white wine
12 gallons (45 lit) of premium red wine
12 gallons (45 lit) of second-run red wine

The recent popularity of white table wines from Cabernet Sauvignon, Pinot noir, Merlot, and Zinfandel has solved the problem of excess red grapes in the vineyards, and the winemaker has the advantage of a wine from red grapes that he can market in 6–8 weeks instead of 1–5 years. Meanwhile, those of us who prefer the complexity of a red finally have a wine that comes from our favorite grapes, yet complements fish and chicken.

If you do not want the labor and expense of working with fresh grapes, there is a winner from Australia — white concentrate from Cabernet Sauvignon.

Recipes using grape concentrates begin on page 134.

FRUIT BASES

Fruit bases consist of tree fruits or soft fruits canned specifically for the winemaker. Surplus apricots, cherries, peaches, and berries too ripe to ship to fresh-fruit markets are processed, so that winemakers receive perfectly ripened fruit at a modest price.

Clearly, fruit bases require the addition of a lot of water and sugar to a small amount of fruit to reduce the fruit flavor and make wine that tastes like fermented vinifera grapes. The addition of grape concentrate makes up for the lack of body that diluted fruit bases produce.

Fruit base wines are not as popular as they used to be, because marketing boards and government controls have increased the price of fruit to the point where grape concentrate, which produces true wine, is just as economical.

Apricot is the most popular of the fruit wine bases, but blackberry, blackcurrant, and cherry make good reds. Acceptable whites are made from peach, gooseberry, apple, and pear.

When using fruit bases, it is important to follow the recipes to the letter. You need to be a very knowledgeable winemaker to alter them without producing a wine with an overwhelming fruit flavor.

Fruit bases have been designed for winemaking only. They contain no preservatives so as to ferment readily, and they contain enough stems and pits to impart an interesting flavor. They are not an economical source of fruit for jams, jellies, or fruit sauces.

Recipes using fruit bases begin on page 31.

Basic Additives

YEAST

Most wineries in the world no longer rely on wild yeast, but add selected strains of wine yeast to ensure a healthy fermentation and a sound wine. The preferred wine yeasts are *Saccharomyces cerevisiae* and *Saccharomyces bayanus.* The commercial producers of yeast either name the various strains for the areas they come from — Champagne, Montrachet, Narbonne, and so on — or they number them. Hundreds of strains have been isolated and propagated by microbiologists, most of them selected from the wine-growing areas of France and Germany.

The ideal yeast will tolerate a wide range of temperatures and a high concentration of alcohol; it will not create bad smells or tastes; and it will clear easily from the wine and leave a firm deposit on the bottom of the fermentor.

Temperature Tolerance

Wine yeasts have been known to grow at temperatures as low as 28°F (-2°C) and will tolerate temperatures up to 110°F (43°C), but they really function best between 60°F (15°C) and 80°F (27°C). We must make a distinction here between yeast *growth* and yeast *fermentation*: For the first 24 hours after it is added to the fermentor, the yeast grows and multiplies. Actual fermentation (alcohol production) starts when yeast cells have multiplied to the optimum population. During the period of growth, called the lag period, the temperature should be 70°F (21°C) or more. Once the population is established — usually within 24 hours — then fermentation can continue at lower temperatures if the winemaker desires. Yeast is somewhat like grass seed; you are unlikely to add too much, but you can add too little. Adding too little means too long a lag period and gives spoilage bacteria a better chance to multiply.

The cooler the temperature, the longer a ferment takes. In Acapulco, a fermentation could be finished within 3 days; in Vancouver, outdoors in the winter, it could last 3 months. Temperatures above 90°F (33°C) will weaken most yeasts and may bring fermentation to a standstill.

Alcohol Tolerance

Table wines contain 10–13% alcohol, but most wine yeasts can be coaxed to produce much more before dying. Yeasts that are fermented at a relatively cool temperature of 60–70°F (15–21°C) and syrup fed (fed small quantities of sugar in liquid form) will produce 18–20% alcohol. Syrup feeding is necessary because although yeast thrives in 22% sugar, large quantities added all at once will bring fermentation to a halt. Sherry yeast fed in this manner under laboratory conditions has produced 23% alcohol by volume.

Deposit and Nose

When the yeast has done its job, it must be removed from the wine. A yeast that flocculates quickly and forms a firm deposit on the bottom of the fermentor allows the wine to be drawn off cleanly.

A good wine yeast must also have what is known as a clean nose. Yeasts produce not only alcohol and carbon dioxide, but also succinic and acetic acid, sulphur dioxide, hydrogen sulphide, and a host of other products in minute quantities. Two products you do not want are hydrogen sulphide, which imparts a smell of bad eggs, and acetic acid, which is vinegar. A yeast free of unpleasant tastes and smells is said to have a good nose.

Which Yeast?

Despite the enormous variety of wines made by modern wineries, they generally use only one or two strains of yeast. In the past we have been guilty of recommending dozens of varieties, but today, after numerous laboratory tests, we feel that 5–6 strains are ample. Although pure strains clearly contribute to the ingredients and the resulting product, a good yeast will not make poor grapes into good wine.

Yeasts are available in a variety of forms:

Colonies of yeast cells on agar slants in test tubes
Liquid cultures in sealed vials
Hydrolized yeast packed in foil envelopes under nitrogen
Hydrolized bulk yeast packed in plastic bags
Sporulated yeast in small paper sachets.

The sealed vials and paper sachets offer the most varieties, some of which are listed below:

Liebfraumilch	Madeira
Rhine	Niersteiner
Riesling	Moselle
Sauterne	Montrachet
Mead	Champagne
Burgundy	Pommard
Bordeaux	Beaujolais
Cabernet	Gamay
Claret	Sherry

For ease of use and reliability, however, we prefer yeasts packed in foil envelopes under nitrogen. Universal Foods of Milwaukee made a major breakthrough when it introduced Montrachet dried wine yeast in the late 1960s. It followed this with champagne and then sherry. Wineries took to using these yeasts very quickly, but disenchantment set in when they realized that Montrachet was liable to develop hydrogen sulphide in the must. We became aware of the problem too and, like the wineries, switched to the champagne strain, which isn't as likely to produce hydrogen sulphide.

Recently, a Montreal company brought out a series of state-of-the-art yeasts under the Lalvin label, also packed in foil under nitrogen for long shelf life. There are 3 currently available:

Montpellier: A vigorous fermenter that tolerates warm temperatures and is recommended for red oxidative wines.

Narbonne: Recommended for fruit wines or reductive wines in which a fruity nose is desired.

Bayanus champagne: Recommended for all white wines.

Montpellier and Bayanus are classified as "killer yeasts"; they are said to exude a protein that kills any rival strain of yeast.

We have no hesitation in recommending these yeasts. They come in 5-gram packages sufficient for 6 gallons (23 lit) of must. At the time of writing they cost about a dollar a package. They do not need culturing before use.

Using Multiple Strains

Some researchers claim that using 2–3 strains of yeast produces better results. State-of-the-art knowledge certainly indicates that adding cultured wine yeast produces the best fermentation, but in our experience, whether you add 1 strain or 3 makes little difference.

For instructions on how to use yeast, see "Yeast Starters," page 240.

SUGARS

Glucose Solids

Dry glucose derived from hydrolized corn starch has recently become available to home winemakers. It is a useful additive when you want to reduce the harshness of a totally dry wine, and it is often used in quick wine recipes. Because it is not entirely fermentable, it gives body and fullness to a wine. It is best used to make up only part of the sugar requirement.

Dextrose

Dextrose is commonly sold under the name *corn sugar*. We originally introduced this to the trade because liquid invert sugar was expensive and difficult to market. Like liquid invert sugar, it is a monosaccharide — a simple form of sugar that yeast does not need to break down to consume.

Enologists argue about the virtues of using invert sugar instead of regular sucrose from cane and beet. Though we cannot prove it scientifically, we believe that better results are obtained by using dextrose. However, dextrose is more expensive, so you would do best to judge for yourself. Try each in a similar recipe and compare the results. All our recipes are for sucrose; so if you decide to use dextrose, remember to use 20% more by weight.

Sucrose

Derived from cane and beet, sucrose is the familiar sugar we use every day in tea, coffee, and baking. It is used in the manufacture of rum and in many wines, especially French and German wines.

Liquid Invert Sugar

Liquid invert is a cane or beet sugar that has been inverted by the application of acid and heat into 2 monosaccharides called dextrose and levulose.

Levulose is much sweeter to the taste than dextrose, but some yeasts do not consume it as easily. For example, Sauterne wine yeast tends to reject the levulose that occurs naturally in the Sauterne grape and by preference easily consumes the natural dextrose, thus leaving in the finished wine a sweet, heady, luscious taste created by the residue of natural levulose.

TANNIN

Tannins — phenolic compounds extracted from the skins, stems, and seeds of grapes — play an important role in wine production. They also contribute significantly to the taste of wine, especially red wine. Most people are familiar with the taste of tannin; it is present in large quantities in black tea and is commonly described as astringent. Wine without tannin is flat and insipid.

Tannin is a natural clearing and fining agent in wine. Suspended proteins combine with tannin to form heavy solids that sink to the bottom of the container. This process is called flocculation. Often, all a cloudy wine needs to make it clear is the addition of tannin. Be sure to use wine-grade tannin, available from winemaking stores. Do not use black tea; you may taste the tea in your wine.

Many home winemakers want their wine to mature fairly quickly; but if you practice the art long enough, you will want to make some fine wines to reserve for special occasions. Here tannin is important because it gives a wine keeping qualities and allows a good red to develop a truly big taste over 5–10 years.

You will find the exact amount of tannin you need specified in each recipe. But don't worry about having too much — it's easy to remove with fining. We prefer to use liquid tannin if it is available. Powdered tannin is difficult to dissolve in water, and we suspect that many winemakers who sprinkle it dry on top of the must get only half the needed amount in solution.

FRUIT ACIDS

The fruit acids used in winemaking — tartaric, malic, and citric — are found in most fruit, including vinifera grapes. Oxidative wines need a higher content of tartaric acid, and reductive wines need more malic acid. There is actually very little citric acid in grapes, but it still has a place in winemaking.

Tartaric Acid

Tartaric acid is the principal acid found in vinifera grapes. If you do not have enough tartaric acid in your wine, it tastes insipid. In some cases, low-acid wines are also bitter and medicinal. We have known winemakers to mistake that bitterness for too much acid.

As fermentation makes wine less dense, the potassium salt in tartaric acid tends to drop out (precipitate) and appear in the container as crystals. These are often referred to as wine stones. Further crystals will drop out if the wine is chillproofed (refrigerated). Sometimes, when you open a well-aged red wine, you will find red crystals stuck to the cork or the side of the bottle. This should not be considered a defect; it usually means that the wine will be soft and lush.

When adjusting the acidity of freshly crushed grapes, winemakers need to take into account this potential loss of tartaric acid through precipitation.

Malic Acid

Malic acid is the second most important acid in grapes. Underripe grapes have a higher content than ripe grapes; but as the sugar increases, the malic acid diminishes. White wines grown in northern climates, such as Germany and New York state, tend to be high in malic acid. They are reductive wines, fruity and fragrant with the aroma of the grape. Often you detect the scent of apples, which contain mostly malic acid.

High malic acid content is considered a defect in red wine, especially an oxidative red wine. In marginal grape-growing areas, such as Canada and the northern U.S., many red grapes have to be harvested with an excessive malic acid content. Fortunately for winemakers, malic acid is subject to attack by a strain of bacteria that converts it into lactic acid. Lactic acid is a much softer-tasting acid and therefore preferable to the palate. In many cases, where wine is stored in a warm cellar and the sulphur dioxide level is low, a malolactic acid ferment occurs naturally during or after the secondary fermentation. In fact, some California enologists claim all their red wines go through this conversion. In northern areas, however, where it is needed most, you cannot rely on a malolactic acid ferment to occur naturally, but you can purchase a culture of leuconostic bacteria and inoculate the wine in the secondary fermentor. (See "Leuconostic Bacteria," page 24.)

Although a malolactic acid fermentation is beneficial during a wine's secondary ferment, we do not consider it desirable once a wine has been bottled. It will not cause the bottle to explode, but the wine will appear bubbly and unstable, and will have an unpleasant nose. Fifty to 100 ppm of sulphite will prevent a malolactic acid ferment, as will a low storage temperature.

Citric Acid

Citric acid is the acid we are all familiar with in lemons, oranges, and grapefruit. Although only a small amount occurs naturally in grapes, wineries like to add some to wines because it lowers the pH and increases the titratable acidity, but does not drop out as crystals. The two drawbacks of citric acid are its potential for breaking down into acetic acid (vinegar) and the fact that some people find citric acid an off taste in wine.

Citric acid is usually the cheapest of the three fruit acids; so when you buy cheap acid blends, you most often get a higher content of citric acid. It is a useful additive, but only in small quantities; so this is not the best way to save money in winemaking.

Vinacid and Acid Blend

Many of our recipes call for the use of Vinacid. This is the trade name for a blend of tartaric, malic, and citric acids premixed for winemakers. Vinacid O is for oxidative wines; Vinacid R is for reductive wines. *Acid blend*, also originally a product devised by us, has now simply become a generic name for blended acid.

SULPHUR DIOXIDE

Sulphur dioxide is produced naturally by wine yeast. It inhibits the growth of spoilage organisms and has a powerful antioxidant effect. Winemakers routinely protect their wine from spoilage by providing additional sulphur dioxide in the form of potassium or sodium metabisulphite. These two salts, used interchangeably and commonly called sulphite, release sulphur dioxide when dissolved in the natural acid solution afforded by wine.* Sulphite is well tolerated by wine yeast, but prevents the growth of organisms such as vinegar bacteria. Wiping with sulphite solution is an effective way to sanitize your equipment and containers before use.

Campden tablets are the easiest way to add a measured amount of sulphite to your wine. Each tablet contains ½ gram of sulphite crystals in a binder and adds approximately 60 ppm to a gallon of must or wine. Straight sulphite crystals are cheaper, and this is why we use them in our recipes, but they require greater care in measuring an exact quantity. There is little danger of toxicity, but if you add too much, your must becomes difficult to ferment.

For instructions in the use of sulphite, see "Sulphite," page 238.

We insist that all our imported concentrates contain sulphur dioxide to protect them from bacteria and oxidation. When you purchase concentrate, be sure to read the manufacturer's label.

STABILIZER

Potassium Sorbate

Potassium sorbate (sorbic acid) is an approved stabilizing additive for wines. When you

*Technologists in the Canadian Department of Consumer Affairs recently pointed out that sulphite in an aqueous solution releases sulphurous acid, not sulphur dioxide. This is technically correct, but we have no intention of trying to change the accepted terminology of the food and wine industry around the world.

sweeten a finished wine, there is always the chance that the added sugar will start a new fermentation in the bottle. Sorbate prevents renewed fermentation by preventing yeast cell budding. It does not kill yeasts or any other organism, however, and it should be used in conjunction with sulphite in a clear, preferably filtered, wine. The minimum used by commercial wineries is 125 ppm; but for home winemakers, 200 ppm in conjunction with 30 ppm of sulphur dioxide is desirable.

Sorbate is approved for use in doses of up to 1,000 ppm and is perfectly harmless in the concentrations winemakers are likely to use. But some people are sensitive to the smell and taste of sorbate. Also, there appear to be undesirable chemical changes in sorbate-treated wines that are aged too long. This is not normally a problem, because the only wines you are likely to sweeten are white reductive wines, which are consumed within 3 months of bottling.

Wine Conditioner

Wine conditioner is a product we devised 20 years ago. By combining sugar and potassium sorbate, it allows you to sweeten finished wines without causing renewed fermentation. The term *wine conditioner* has now entered general usage, and several manufacturers now offer a similar product. Our formula uses a 77 Brix liquid sugar that is 50% invert. The extra-sweet, high-density liquid permits sweetening of the wine without too much dilution of its alcohol concentration. It contains enough sorbate so that 2 ounces (60 ml) will slightly sweeten 1 gallon (4 lit) of wine without risk of fermentation. The wine should contain at least 10% alcohol, and 30 ppm of sulphur dioxide.

The restrictions in the use of sorbate also apply to wine conditioner; the product is therefore recommended only for reductive wines that are intended for consumption within 6 months of bottling.

ENZYMES

Enzymes are organic substances that act as catalysts; that is, they produce changes in substances they come in contact with, but do not change themselves. Many enzymes occur naturally in plants and animals, but here we are interested only in those used in winemaking.

Pectinase

Pectin is the substance in fruit that makes jams and jellies set. Most people add it to their jams in the form of Certo, which is a brand name for pectin. In wines, however, it is undesirable because it causes the formation of unsightly jelly-like lumps. Pectinase breaks down the pectins and they settle out, leaving your wine clear.

Failing to add pectinase won't spoil the taste of your wine, but after fermentation the wine may remain cloudy, and all your attempts to fine or filter it will not solve the problem. We have had winemakers phone us with the complaint that a year after fermentation the blackberries were forming again in their blackberry wine. They thought that a miracle was occurring. It wasn't the blackberries; it was blobs of pectin. Pectinase made that miracle disappear overnight!

Pectic enzymes come in powder or liquid form. Powder keeps longer, but all pectinase loses strength during storage, so try to buy what you need and use it while it's fresh — within 3 months. There are several strengths of pectinase, so use the quantity suggested on the package.

Pectinase is used by most fruit processors because it improves color and increases the release of juice by up to 17%. It is used with all fruits from which the end product required is clear juice, such as apples, pears, grapes, blackberries, and so on.

Most forms of pectinase work well between 70°F (21°C) and 140°F (60°C), but are destroyed at 149°F (65°C). The warmer the must or fruit (within that range), the faster the enzyme works. While there is relatively little pectin in ripe vinifera grapes, many wineries do add pectinase for greater juice yield and clearer wines. We recommend it for most wines.

Diastase

You will find diastase mentioned in winemaking books, but we question the need for it in winemaking. It converts starch to sugar, but unless you make wine from potatoes, parsnips, or grains, you should not need diastase.

Proteinase

Proteinase, too, is occasionally recommended. As the name implies, it breaks down proteins. It can be useful to beermakers, who sometimes have beer that goes cloudy in the refrigerator, but we have never seen protein haze in wine.

ASCORBIC ACID (VITAMIN C)

The most popular vitamin of the last 20 years among the health-conscious, vitamin C also has uses in the food industry because of its antioxidant effect. As we have observed, oxidation means deterioration — as when an apple turns brown, or a steel tower rusts. When wine oxidizes, it is spoiled; oxidation is therefore a major problem for the winemaker.

Ascorbic acid is not as effective an antioxidant as sulphur dioxide (SO_2), but there are limits to the amount of SO_2 we can use; so we turn to ascorbic acid to do part of the job. In fact, ascorbic acid and sulphur dioxide used together are synergistic — they aid and abet each other so you can use less of both. By all means, use vitamin C to help your sulphur dioxide work better, but do not try to replace SO_2 with ascorbic acid only. Some suppliers use sodium erythorbate instead of ascorbic. This is an isomer (mirror image) of ascorbic acid. It works well as an antioxidant, but does not have the therapeutic value of vitamin C.

CHALK AND OTHER ANTACIDS

In northern climates, even vinifera grapes are often high in acid. To avoid making a thin wine by adding water to reduce the acid, many winemakers add an alkaline substance, such as chalk, to neutralize some of the acid.

The most popular product is called Acidex. The winery procedure is to take a portion of the juice prior to the onset of fermentation, run it into a tank with an agitator, and add enough Acidex to remove all of the acid. After this deacidified juice is clear — a process taking up to 24 hours — it is then blended back into the bulk of the juice and fermented.

The smaller the proportion of juice you need to neutralize, the better the results. If you can't get Acidex, precipitated chalk, food-grade calcium carbonate, and potassium carbonate are also effective. To use Acidex or chalk, see "Lowering the Acid Level," page 222.

Our favorite procedure with high-acid grapes, however, is to blend them with a low-acid, good-quality grape concentrate of a compatible variety. (See "Hybrid Grape Use," page 262.)

OAK

Oak makes a great contribution to the flavor of red table wine. If you do not use oak barrels, it really does help to use an oak derivative as a substitute, and there are two or three products available.

Some of our recipes call for a popular oak extract from Spain called Sinatin 17. It makes the wine smell and taste more vinous in a way that we associate with mature wines, and it appears to hasten the wine's aging. There are other brands of oak extract on the market but Sinatin 17 is the only one we have widely tested.

Oak Chips and Oak Sticks

Oak chips and oak sticks from French oak coopers are another excellent way to add oak flavor to your wine, but their effect is more difficult to calculate. Worse than no oak in your wine is too much; and once the flavor is in, you can't get it out. We remember when Robert Mondavi first fell in love with oak in red wine — the odor of oak in his wine was all you could smell. We're glad he got over that infatuation, because we love his wine now.

To use oak chips, soak them in a standard sulphite solution first for a few minutes. This will remove the dust and prevent explosive foaming when you syphon the wine over the chips in the bottom of your carboy. Do not leave oak chips in your wine too long. After a week, start smelling the wine, and if you think you smell oak, that's enough. Rack the wine off the oak chips into another carboy.

Oak sticks are about 8 inches by ¾ inch (20 x 2 cm) and are perhaps a more convenient way to add oak flavor to wines in carboys. Cutting them in half makes them a little more economical. Soak them in a standard sulphite solution before inserting them in the wine. Again, check the aroma of your wine often to avoid overdoing it.

GLYCERINE

Glycerine is often used as a sweetener for wine. It is somewhat underused by winemakers — perhaps because we haven't included it in enough recipes. It will not ferment, is nontoxic, and is thick and syrupy enough to add body to wine. If you can't get wine conditioner, or you want only a touch of added sweetness to your wine, try 1 ounce (30 ml) of glycerine per gallon (4 lit). It's particularly good in sherries and ports, where body is very important.

LEUCONOSTIC BACTERIA

Leuconostic bacteria convert malic acid, which is tart, to lactic acid, which is milder-tasting. The presence of leuconostic bacteria can therefore improve wines high in malic acid, such as those made from hybrid grapes. In warm climates, inoculation usually takes place naturally during secondary fermentation. In cooler climates, where it is needed most, a leuconostic bacteria culture must usually be added. Cultures are not readily available to amateur winemakers, but one outlet, Wine Lab,* supplies a liquid culture that is added directly to the wine at 70°F (21°C) (20 ppm SO_2).

ELDERFLOWERS

Dried elderflowers impart a Riesling-like bouquet to white wine. Use a small amount, 1 ounce (30 g) per 5 gallons (19 lit). Too much is worse than none.

BANANA FLAKES

Dried banana powder or banana flakes add body without adding flavor to aperitif and dessert wines. Do not use more than 1 ounce (30 g) per gallon (4 lit).

BILBERRIES

Dried bilberries come from Europe and are similar to blueberries. They add body and tannin to red wine that might otherwise lack vinous qualities.

SMOOTHY CRYSTALS

Smoothy Crystals is a trade name for granular glucose used to add body or thickness to liqueurs without adding a sickly-sweet flavor. They are also added to light table wines; the result is a wine with more body and softness.

*See "Buyer's Guide," page 281.

PART TWO

Recipes

How to Use the Recipes

Alongside each recipe step that requires knowledge of a winemaking technique you will find a symbol as follows:

- Measuring specific gravity
- Measuring temperature
- Starting yeast
- Racking
- Attaching a fermentation lock
- Adding finings
- Measuring the acid level
- Filtering
- Adding sulphite
- Bottling
- Aging

To find instructions on these techniques, look up the "Techniques" page reference given at the foot of the page.

HOW WE CALCULATE OUR RECIPES

All our recipes are for 5-gallon (19 lit) batches. We specify less water at the outset so that you can top up with tap water at each racking without upsetting the acid and alcohol balance. Many winemakers seek to avoid dilution by setting aside small leftover quantities of wine to be used exclusively for topping up, but in so doing they often contribute unwittingly to the spoilage of their wines. Small quantities kept in imperfectly sealed containers are especially vulnerable to bacteria and oxidation. When they are introduced to the carboy as a top-up, they quickly lead to the infection of the whole batch.

We have calculated the recipes so that you can dilute the wine with risk-free tap water at each racking and still arrive at a finished wine with a perfect acid and alcohol content. Naturally, this will mean specific gravity and acid readings that are unusually high at the start. Do not be misled; a starting SG of 1.100 and an acid content of 5 g/l, when topped up at 3 rackings, will give you a finished wine with 12.3% alcohol and a total acid of 6.3 g/l — a well-balanced table wine.

Topping Up Fresh Grape Wines

Because our fresh grape recipes involve significantly larger volumes of wine, and because fresh grapes themselves produce an unpredictable quantity of lees, it is difficult to calculate fresh grape recipes to fit a specific container size, and difficult to predict the volumes of wine remaining after each racking. Consequently, when you top up fresh grape wines, you will be adding quite large volumes of liquid to your secondary fermentors. Topping up a 50-gallon barrel for one year, for example, will need at least 3–5 gallons of liquid. Obviously, such large quantities of tap water would significantly impair the quality of the wine. With fresh grape wines, therefore, keeping quantities of top-up wine on hand becomes essential. You will find instructions on keeping low-risk top-up wine in "Quick Reference for Fresh Grape Wines," page 181.

MEASUREMENTS AND CONVERSIONS

All conversions in this book are from U.S. standard measure to metric measure. (There is no reference to imperial measure in any conversion.)

To eliminate fractions and decimal points, we have rounded off some conversions where the difference will not affect the outcome; for instance, 1 U.S. gallon equals 3.7854 liters; but we, like the packaging industry, call it 4 liters. Similarly, we represent a 96-ounce can of concentrate as 3 liters metric, and a 140-ounce can of concentrate (Italian or French packaging) as 4 liters metric.

BEFORE YOU BEGIN

Choose an area in your home to place the fermentor. Primary fermentation needs a temperature of 75°F (23°C). Choose a place that will allow you to maintain this temperature for 5 days — with a heating pad or belt if necessary. If you live in a small apartment and your space is limited, a hall closet or a corner hidden behind a decorator screen works fine.

Because you will be syphoning from the primary fermentor, you should place it approximately 3 feet above the ground, on a secure stool or chair. When full, the fermentor will weigh 60–70 pounds (27–32 kg), so adjusting its position will be difficult after fermentation has begun. You can purchase a small platform on castors from any winemaking retail outlet, or you can make one by attaching sturdy castors to an 18- by 18-inch (46- by 46-cm) square of ½-inch plywood. This size allows you enough room to keep your hydrometer and recipe book within reach. A movable platform also gives you the freedom to move the fermentor to a warmer or cooler location as necessary.

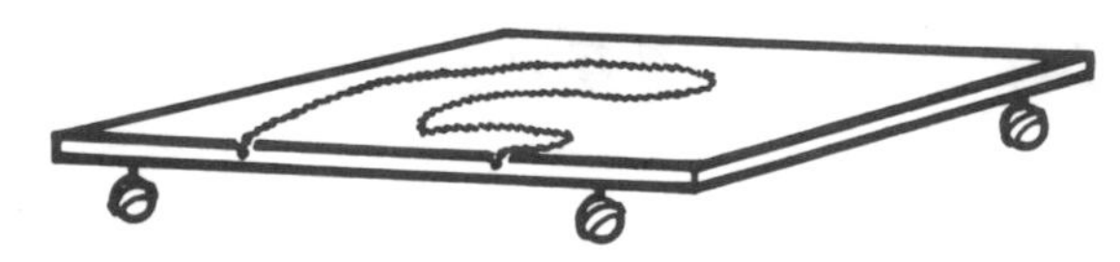

Do-it-yourself platform for the primary fermentor

Quick Wine

If you want drinkable wine within 28 days, buy a quick-wine kit. Most winemaking supply stores offer a range to choose from. Select the one with the largest volume of grape concentrate. Avoid kits that provide acid crystals instead of concentrate; with sugar and water they may ferment, but the result is not wine. Start with a white wine; you can make an acceptable white table wine in 3–5 weeks.

We recommend a 5-gallon (19–23 lit) kit, which will provide you with 25–30 bottles of wine. You will need

Quick-wine kit
Food-grade plastic tub
Plastic carboy
Fermentation lock and bung to fit the carboy
Plastic sheet
Glass jugs with screw caps to hold 5 gallons
Syphon hose

BERRY WINES

QUICK REFERENCE FOR BERRY WINES

Washing

If fruit has been sprayed recently with insect repellent or foliant, wash it before processing.

Fruit Quality

Do not use overripe, moldy, or fermenting fruit.

Fresh or Frozen?

You may use fresh or frozen fruit. Freezing improves fruit for winemaking because it breaks down the cell structure and allows juice to be released more easily. To freeze fruit until a time convenient for winemaking, sprinkle it with a sulphite solution (3 Campden tablets or ¼ teaspoon of sulphite crystals per 2 cups of water), then freeze it in 5 lb (2.2 kg) freezer bags. Note: When using fruit that has been sulphited before freezing, you do not need to add the sulphite listed in the primary recipe ingredients.

Crushing

To crush berries, use a stainless-steel potato masher, a wooden mallet, or the bottom of a heavy glass preserving jar. Do not use a blender, or you will release a bitter taste from the broken seeds. Do not use a steamer to extract juices, because steaming oxidizes juice.

Pressing

To press berries, place the fruit in a nylon straining bag and squeeze the juice out by hand or place the fruit in a colander and press the juice out with a mallet or a wooden spoon.

Used Pulp

Discard used berry pulp. It cannot be used for a second run.

Blackberry Bordeaux

READY: 1 YEAR

Well made and well aged, this blackberry table wine resembles a Bordeaux.

PRIMARY INGREDIENTS

20 lb	Blackberries (crushed)	9 kg
1 qt	Red grape concentrate	1 lit
9 lb	Sugar	4 kg
6 qt	HOT water	6 lit
2 tsp	Yeast nutrient	
2 tsp	Pectic enzyme	
2 tsp	Liquid tannin	
8	Campden tablets (crushed)	
8 qt	COLD water	8 lit
1 pkt	Montpellier wine yeast	

SECONDARY INGREDIENTS

	Gelatin finings	
¼ tsp	Sulphite crystals	
1 oz	Sinatin 17	30 ml

EQUIPMENT

Basic 10 + straining bag and potato masher; coarse filter pads

PRIMARY SEQUENCE

1. Crush the blackberries and place them with the grape concentrate in the primary fermentor. Add the hot water and sugar.

2. Stir thoroughly until all the sugar is dissolved.

3. Add the next 5 ingredients. Mix well.

4. Check, and if necessary adjust, the specific gravity (SG) of the must. It should be 1.100.

5. Check, and if necessary adjust, the temperature of the must. It should be 75°F (23°C).

6. Add the yeast to a cup of warm water. Let stand for 10 minutes. Stir in.

7. Cover the fermentor with a plastic sheet; tie down. Keep in a warm place (75°F [23°C]). After 24 hours, check that fermentation has begun. Foam should be visible on the surface and/or bubbles should be audible. If fermentation has not begun, see "Stuck Ferment," page 269.

8. Stir twice daily to keep the floating fruit moist.

9. Check the SG every other day.

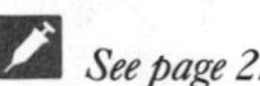
See page 220

See page 223
See page 224

See page 226
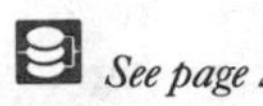
See page 227

See page 229

SECONDARY SEQUENCE

10. When SG reaches 1.020, scoop the blackberries into a straining bag and squeeze the juice gently into the fermentor. Discard the pulp.

11. Rack into a clean carboy. Top up with cold tap water.

12. Attach the fermentation lock.

13. Move to a cooler location, ideally 65°F (18°C).

14. After 10 days or at SG 1.000, whichever comes first, rack into a clean carboy. Top up with cold tap water.

15. After 3 weeks or at SG .990–.995, whichever comes first, rack into a clean carboy.

16. Add the finings. Top up with cold tap water. Let rest 10 days.

17. Rack into the primary fermentor.

18. Filter into a clean carboy.

19. Add ¼ teaspoon of sulphite crystals dissolved in a small amount of water. Add the Sinatin 17. Top up with cold tap water.

20. Bulk age 3 months.

21. Bottle.

22. Bottle age 9 months.

See page 232 (col. 1) *See page 232 (col. 2)*  *See page 238*  *See page 240 (col. 1)* 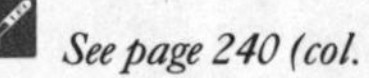  *See page 240 (col. 2)*

Blackberry Social

READY: 7 MONTHS

A fruit wine that matures quickly and is slightly sweet

PRIMARY INGREDIENTS

24 lb	Blackberries (crushed)	10.9 kg
13 lb	Sugar	6 kg
6 qt	HOT water	6 lit
2 tsp	Yeast nutrient	
2 tsp	Pectic enzyme	
8	Campden tablets (crushed)	
8 qt	COLD water	8 lit
1 pkt	Narbonne wine yeast	

SECONDARY INGREDIENTS

	Bentonite finings	
¼ tsp	Sulphite crystals	
10 oz	Wine conditioner	300 ml

EQUIPMENT

Basic 10 + straining bag and potato masher; coarse filter pads

PRIMARY SEQUENCE

1. Crush the blackberries and place them in the primary fermentor. Add the hot water and sugar.

2. Stir thoroughly until all the sugar is dissolved.

3. Add the next 4 ingredients. Mix well.

4. Check, and if necessary adjust, the temperature of the must. It should be 75°F (23°C).

5. Add the yeast to a cup of warm water. Let stand for 10 minutes. Stir in.

6. Cover the fermentor with a plastic sheet; tie down. Keep in a warm place (75°F [23°C]). After 24 hours, check that fermentation has begun. Foam should be visible on the surface and/or bubbles should be audible. If fermentation has not begun, see "Stuck Ferment," page 269.

7. Stir twice daily to keep the floating fruit moist.

8. Check the specific gravity (SG) every other day.

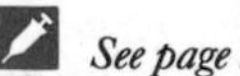 See page 220   See page 223  See page 224  See page 226  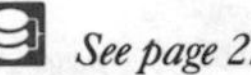 See page 227  See page 229

SECONDARY SEQUENCE

9. When SG reaches 1.020, scoop the blackberries into a straining bag and squeeze the juice out gently into the fermentor. Discard the pulp.

10. Rack into a clean carboy. Top up with cold tap water.

11. Attach the fermentation lock.

12. Move to a cooler location, ideally 65°F (18°C).

13. After 10 days or at SG 1.000, whichever comes first, rack into a clean carboy. Top up with cold tap water.

14. After 3 weeks or at SG .990–.995, whichever comes first, rack into a clean carboy.

15. Add the finings. Top up with cold tap water. Let rest 10 days.

16. Rack into the primary fermentor.

17. Filter into a clean carboy.

18. Add ¼ teaspoon of sulphite crystals dissolved in a small amount of water. Top up with cold tap water.

19. Bulk age 2 months.

20. Add the wine conditioner and bottle.

21. Bottle age 4 months.

 *See page 232 (col. 1)*  *See page 232 (col. 2)*  *See page 238* 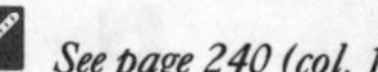 *See page 240 (col. 1)*  *See page 240 (col. 2)*

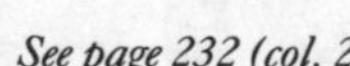

Port-Style Blackberry Wine

READY: 1 YEAR

Delicious served as a dessert wine with cheese or fruit

PRIMARY INGREDIENTS

20 lb	Blackberries (crushed)	9 kg
8 oz	Dried elderberries	240 g
8 oz	Banana powder	240 g
13 lb	Sugar	6 kg
6 qt	HOT water	6 lit
2 tsp	Yeast nutrient	
2 tsp	Pectic enzyme	
8	Campden tablets (crushed)	
8 qt	COLD water	8 lit
1 pkt	Wine yeast with a high alcohol tolerance	

SECONDARY INGREDIENTS

	Gelatin finings	
4 oz	Glycerine	120 ml
1 oz	Sinatin 17	30 ml
12 oz	Brandy	360 ml
¼ tsp	Sulphite crystals	
10 oz	Wine conditioner	300 ml

(continues)

EQUIPMENT

Basic 10 + straining bag and potato masher; coarse filter pads

PRIMARY SEQUENCE

1. Crush the blackberries and place them in the primary fermentor. Add the elderberries, banana powder, sugar, and hot water.

2. Stir thoroughly until all the sugar is dissolved.

3. Add the next 4 ingredients. Mix well.

4. Check, and if necessary adjust, the temperature of the must. It should be 75°F (23°C).

5. Add the yeast to a cup of warm water. Let stand for 10 minutes. Stir in.

6. Cover the fermentor with a plastic sheet; tie down. Keep in a warm place (75°F [23°C]).

See page 220

 *See page 223*

 *See page 224*

See page 226

 *See page 227*

 *See page 229*

After 24 hours, check that fermentation has begun. Foam should be visible on the surface and/or bubbles should be audible. If fermentation has not begun, see "Stuck Ferment," page 269.

7. Stir twice daily to keep the floating fruit moist.

8. Check the specific gravity (SG) every other day.

SECONDARY SEQUENCE

9. When SG reaches 1.020, scoop the fruit into a straining bag and squeeze the juice gently into the fermentor. Discard the pulp.

10. Rack into a clean carboy. Top up with cold tap water.

11. Attach the fermentation lock.

12. Move to a cooler location, ideally 65°F (18°C).

13. After 10 days or at SG 1.000, whichever comes first, rack into a clean carboy. Top up with cold tap water.

14. After 3 weeks, rack into a clean carboy.

15. Add the finings. Top up with cold tap water. Let rest 10 days.

16. Rack into a clean carboy. Add the Sinatin 17, glycerine, and brandy.

17. Bulk age 3 months.

18. Rack. Add the wine conditioner.

19. Filter into bottles.

20. Bottle age 9 months.

See page 232 (col. 1)  *See page 232 (col. 2)*  *See page 238*  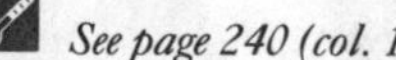 *See page 240 (col. 1)*  *See page 240 (col. 2)*

Blackcurrant Wine

READY: 1 YEAR

A very aromatic wine; quite complex. Deserves aging and could live 5 years. It should be sweetened slightly to bring out its full flavor and aroma.

PRIMARY INGREDIENTS

12 lb	Blackcurrants (crushed)	5.5 kg
1 qt	Red grape concentrate	1 lit
10 lb	Sugar	4.5 kg
6 qt	HOT water	6 lit
2 tsp	Yeast nutrient	
8	Campden tablets (crushed)	
2 tsp	Pectic enzyme	
8 qt	COLD water	8 lit
1 pkt	Narbonne wine yeast	

SECONDARY INGREDIENTS

	Claro K. C. finings	
8 oz	Wine conditioner	240 ml
¼ tsp	Sulphite crystals	

EQUIPMENT

Basic 10 + straining bag and potato masher; coarse filter pads

PRIMARY SEQUENCE

1. Crush the blackcurrants and place them in the primary fermentor. Add the concentrate, sugar, and hot water.

2. Stir thoroughly until all the sugar is dissolved.

3. Add the next 4 ingredients. Mix well.

4. Check, and if necessary adjust, the temperature of the must. It should be 75°F (23°C).

5. Add the yeast to a cup of warm water. Let stand for 10 minutes. Stir in.

6. Cover the fermentor with a plastic sheet; tie down. Keep in a warm place (75°F [23°C]). After 24 hours, check that fermentation has begun. Foam should be visible on the surface and/or bubbles should be audible. If fermentation has not begun, see "Stuck Ferment," page 269.

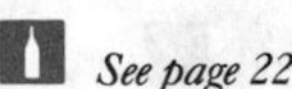

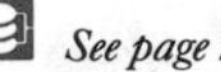

See page 220 *See page 223* *See page 224* *See page 226* *See page 227* *See page 229*

7. Stir twice daily to keep the floating fruit moist.

8. Check the specific gravity (SG) every other day.

SECONDARY SEQUENCE

9. When SG reaches 1.020, scoop the blackcurrants into a straining bag and squeeze as dry as possible into the fermentor. Discard the pulp.

10. Rack into a clean carboy. Top up with cold tap water.

11. Attach the fermentation lock.

12. Move to a cooler location, ideally 65°F (18°C).

13. After 10 days or at SG 1.000, whichever comes first, rack into a clean carboy. Top up with cold tap water.

14. After 3 weeks or at SG .990–.995, whichever comes first, rack into a clean carboy.

15. Add the finings. Top up with cold tap water. Let rest 10 days.

16. Rack into the primary fermentor.

17. Filter into a clean carboy.

18. Add ¼ teaspoon of sulphite crystals dissolved in a small amount of water. Top up with cold tap water.

19. Bulk age 3 months.

20. Rack and add the wine conditioner.

21. Bottle.

22. Bottle age 9 months.

See page 232 (col. 1)

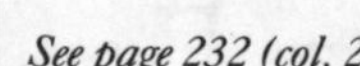
See page 232 (col. 2)

See page 238

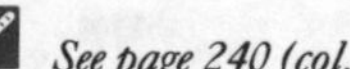
See page 240 (col. 1)

See page 240 (col. 2)

Blueberry Burgundy

READY: 1 YEAR

Compares surprisingly well to a good imported red table wine

PRIMARY INGREDIENTS

Imperial	Ingredient	Metric
12 lb	Blueberries (crushed)	5.5 kg
1 qt	Red grape concentrate	1 lit
9 lb	Sugar	4 kg
10 tsp	Vinacid O	
6 qt	HOT water	6 lit
2 tsp	Yeast nutrient	
2 tsp	Pectic enzyme	
1 tsp	Liquid tannin	
8	Campden tablets (crushed)	
8 qt	COLD water	8 lit
1 pkt	Montpellier wine yeast	

SECONDARY INGREDIENTS

Imperial	Ingredient	Metric
	Claro K. C. finings	
1 oz	Sinatin 17	30 ml
¼ tsp	Sulphite crystals	

EQUIPMENT

Basic 10 + straining bag and potato masher; coarse filter pads

PRIMARY SEQUENCE

1. Crush the blueberries and place them in the primary fermentor. Add the grape concentrate and the hot water, sugar and Vinacid O.

2. Stir thoroughly until all the sugar is dissolved.

3. Add the next 5 ingredients. Mix well.

4. Check, and if necessary adjust, the specific gravity (SG) of the must. It should be 1.105.

5. Check, and if necessary adjust, the temperature of the must. It should be 75°F (23°C).

6. Add the yeast to a cup of warm water. Let stand for 10 minutes. Stir in.

7. Cover the fermentor with a plastic sheet; tie down. Keep in a warm place (75°F [23°C]). After 24 hours, check that fermentation has begun. Foam should be visible on the surface

See page 220 See page 223 See page 224 See page 226 See page 227 See page 229

and/or bubbles should be audible. If fermentation has not begun, see "Stuck Ferment," page 269.

8. Stir twice daily to keep the floating fruit moist.

9. Check SG every other day.

SECONDARY SEQUENCE

10. When SG reaches 1.020, scoop the fruit into a straining bag and squeeze as dry as possible over the fermentor. Discard the pulp.

11. Rack into a clean carboy. Top up with cold tap water.

12. Attach the fermentation lock.

13. Move to a cooler location, ideally 65°F (18°C).

14. After 10 days or at SG 1.000, whichever comes first, rack into a clean carboy. Top up with cold tap water.

15. After 3 weeks or at SG .990–.995, whichever comes first, rack into a clean carboy.

16. Add the finings. Top up with cold tap water. Let rest 10 days.

17. Rack into the primary fermentor.

18. Filter into a clean carboy.

19. Add ¼ teaspoon of sulphite crystals dissolved in a small amount of water. Add the Sinatin 17. Top up with cold tap water.

20. Bulk age 3 months.

21. Bottle.

22. Bottle age 9–12 months.

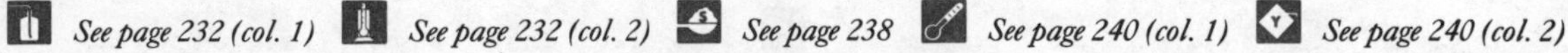
See page 232 (col. 1)

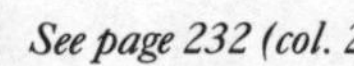
See page 232 (col. 2)

See page 238

See page 240 (col. 1)

See page 240 (col. 2)

Elderberry Wine

READY: 1 YEAR

Famous in England, where the blue-black elderberries are widely available. Properly aged, it resembles a European table wine. In some locations, wild elderberries are bright orange and contain a bitter oil; do not use these for winemaking.

PRIMARY INGREDIENTS

12 lb	Fresh elderberries (crushed)	5.5 kg
1 qt	Red grape concentrate	1 lit
10 lb	Sugar	4.5 kg
6 qt	HOT water	6 lit
2 tsp	Yeast nutrient	
2 tsp	Pectic enzyme	
8	Campden tablets (crushed)	
8 qt	COLD water	8 lit
1 pkt	Montpellier wine yeast	

SECONDARY INGREDIENTS

	Gelatin finings	
¼ tsp	Sulphite crystals	

EQUIPMENT

Basic 10 + straining bag and potato masher; coarse filter pads

PRIMARY SEQUENCE

1. Crush the elderberries and place them in the primary fermentor. Add the concentrate, sugar, and hot water.

2. Stir thoroughly until all the sugar is dissolved.

3. Add the next 4 ingredients. Mix well.

4. Check, and if necessary adjust, the specific gravity (SG) of the must. It should be 1.105.

5. Check, and if necessary adjust, the temperature of the must. It should be 75°F (23°C).

6. Add the yeast to a cup of warm water. Let stand for 10 minutes. Stir in.

7. Cover the fermentor with a plastic sheet; tie down. Keep in a warm place (75°F [23°C]). After 24 hours, check that fermentation has begun. Foam should be visible on the surface

See page 220  *See page 223*  *See page 224*  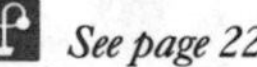 *See page 226*   *See page 227*  *See page 229*

and/or bubbles should be audible. If fermentation has not begun, see "Stuck Ferment," page 269.

8. Stir twice daily to keep the floating fruit moist.

9. Check SG every other day.

SECONDARY SEQUENCE

10. When SG reaches 1.020, scoop the elderberries into a straining bag and squeeze as dry as possible into the fermentor. Discard the pulp.

11. Rack into a clean carboy. Top up with cold tap water.

12. Attach the fermentation lock.

13. Move to a cooler location, ideally 65°F (18°C).

14. After 10 days or at SG 1.000, whichever comes first, rack into a clean carboy. Top up with cold tap water.

15. After 3 weeks or at SG .990–.995, whichever comes first, rack into a clean carboy.

16. Add the finings. Top up with cold tap water. Let rest 10 days.

17. Rack into the primary fermentor.

18. Filter into a clean carboy.

19. Add ¼ teaspoon of sulphite crystals dissolved in a small amount of water. Top up with cold tap water.

20. Bulk age 3 months.

21. Bottle.

22. Bottle age 9 months.

See page 232 (col. 1)  *See page 232 (col. 2)*  *See page 238*  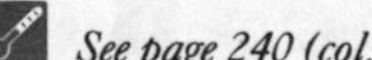 *See page 240 (col. 1)*  *See page 240 (col. 2)*

Loganberry Wine

READY: 8 MONTHS

The Pacific Northwest is famous for its loganberries. They are a high-acid fruit, so you must sweeten the wine to appreciate its full flavor.

PRIMARY INGREDIENTS

12 lb	Fresh or frozen loganberries (crushed)	5.5 kg
11 lb	Sugar	5 kg
6 qt	HOT water	6 lit
2 tsp	Yeast nutrient	
2 tsp	Pectic enzyme	
1 tsp	Liquid tannin	
8	Campden tablets (crushed)	
8 qt	COLD water	8 lit
1 pkt	Narbonne wine yeast	

SECONDARY INGREDIENTS

	Claro K. C. finings	
¼ tsp	Sulphite crystals	
10 oz	Wine conditioner	300 ml

EQUIPMENT

Basic 10 + straining bag and potato masher; fine filter pads

PRIMARY SEQUENCE

1. Crush the loganberries and place them in the primary fermentor. Add the hot water and sugar.

2. Stir thoroughly until all the sugar is dissolved.

3. Add the next 5 ingredients. Mix well.

4. Check, and if necessary adjust, the specific gravity (SG) of the must. It should be 1.100.

5. Check, and if necessary adjust, the temperature of the must. It should be 75°F (23°C).

6. Add the yeast to a cup of warm water. Let stand for 10 minutes. Stir in.

7. Cover the fermentor with a plastic sheet; tie down. Keep in a warm place (75°F [23°C]). After 24 hours, check that fermentation has begun. Foam should be visible on the surface

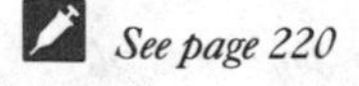 See page 220  See page 223  See page 224  See page 226  See page 227   See page 229

and/or bubbles should be audible. If fermentation has not begun, see "Stuck Ferment," page 269.

8. Stir twice daily to keep the floating fruit moist.

9. Check SG every other day.

SECONDARY SEQUENCE

10. When SG reaches 1.020, scoop the fruit into a straining bag and squeeze the juice gently into the fermentor. Take care not to squeeze the seeds through the mesh. Discard the pulp.

11. Rack into a clean carboy. Top up with cold tap water.

12. Attach the fermentation lock.

13. Move to a cooler location, ideally 65°F (18°C).

14. After 10 days or at SG 1.000, whichever comes first, rack into a clean carboy. Top up with cold tap water.

15. After 3 weeks or at SG .990–.995, whichever comes first, rack into a clean carboy.

16. Add the finings. Top up with cold tap water. Let rest 10 days.

17. Rack into the primary fermentor.

18. Filter into a clean carboy.

19. Add ¼ teaspoon of sulphite crystals dissolved in a small amount of water. Add the wine conditioner. Top up with cold tap water.

20. Bulk age 1 month.

21. Bottle.

22. Bottle age 6 months.

See page 232 (col. 1) *See page 232 (col. 2)* *See page 238* *See page 240 (col. 1)* *See page 240 (col. 2)*

Raspberry Wine

READY: 6 MONTHS

A delicious wine, but it must be slightly sweetened to bring out the flavor and aroma. The berries must be frozen first, then thawed in fine straining bags; this prevents the seeds from spoiling the wine.

PRIMARY INGREDIENTS

15 lb	Frozen raspberries	6.8 kg
10 lb	Sugar	4.5 kg
4 tsp	Vinacid R	
6 qt	HOT water	6 lit
2 tsp	Yeast nutrient	
2 tsp	Liquid tannin	
2 tsp	Pectic enzyme	
8	Campden tablets (crushed)	
8 qt	COLD water	8 lit
1 pkt	Narbonne wine yeast	

SECONDARY INGREDIENTS

	Claro K. C. finings	
¼ tsp	Sulphite crystals	
10 oz	Wine conditioner	300 ml

EQUIPMENT

Basic 10 + fine mesh straining bag and fine filter pads

PRIMARY SEQUENCE

1. When the raspberries have thawed, squeeze the straining bags as dry as possible and collect the juice. Discard the pulp.

2. Place the raspberry juice in the primary fermentor. Add the hot water, sugar, and Vinacid R.

3. Stir thoroughly until all the sugar is dissolved.

4. Add the next 5 ingredients. Mix well.

5. Check, and if necessary adjust, the specific gravity (SG) of the must. It should be 1.100.

6. Check, and if necessary adjust, the temperature of the must. It should be 75°F (23°C).

7. Add the yeast to a cup of warm water. Let stand for 10 minutes. Stir in.

 See page 220  See page 223  See page 224  See page 226 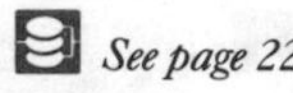 See page 227 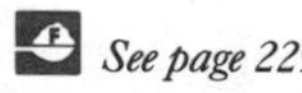 See page 229

8. Cover the fermentor with a plastic sheet; tie down. Keep in a warm place (75°F [23°C]). After 24 hours, check that fermentation has begun. Foam should be visible on the surface and/or bubbles should be audible. If fermentation has not begun, see "Stuck Ferment," page 269.

9. With a colander, remove as much floating pulp as possible.

10. Check SG every other day.

SECONDARY SEQUENCE

11. When SG reaches 1.020, rack into a clean carboy. Top up with cold tap water.

12. Attach the fermentation lock.

13. Move to a cooler location, ideally 65°F (18°C).

14. After 10 days or at SG 1.000, whichever comes first, rack into a clean carboy. Top up with cold tap water.

15. After 3 weeks or at SG .990–.995, whichever comes first, rack into a clean carboy.

16. Add the finings. Top up with cold tap water. Let rest 10 days.

17. Rack into the primary fermentor.

18. Filter into a clean carboy.

19. Add ¼ teaspoon of sulphite crystals dissolved in a small amount of water. Top up with cold tap water.

20. Bulk age 2 months.

21. Add wine conditioner and bottle.

22. Bottle age 3 months.

See page 232 (col. 1)  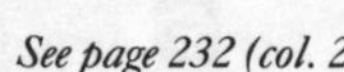 *See page 232 (col. 2)*  *See page 238*  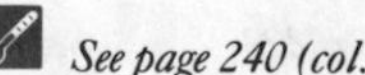 *See page 240 (col. 1)*  *See page 240 (col. 2)*

Saskatoon Berry Wine

READY: 1 YEAR

Saskatoon berries are available only in the Western Prairies, where they grow wild. They make a dry table wine, similar to blackberry wine; but the berries must be fully ripe.

PRIMARY INGREDIENTS

12 lb	Saskatoon berries (crushed)	5.5 kg
1 qt	Red grape concentrate	1 lit
12 lb	Sugar	5.5 kg
10 tsp	Vinacid O	
6 qt	HOT water	6 lit
2 tsp	Yeast nutrient	
2 tsp	Pectic enzyme	
8	Campden tablets (crushed)	
8 qt	COLD water	8 lit
1 pkt	Montpellier wine yeast	

SECONDARY INGREDIENTS

	Claro K. C. finings	
¼ tsp	Sulphite crystals	
1 oz	Sinatin 17	30 ml

(continues)

EQUIPMENT

Basic 10 + straining bag and potato masher; coarse filter pads

PRIMARY SEQUENCE

1. Crush the saskatoon berries and place them, together with the grape concentrate, in the primary fermentor. Add the hot water, sugar, and Vinacid O.

2. Stir thoroughly until all the sugar is dissolved.

3. Add the next 4 ingredients. Mix well.

4. Check, and if necessary adjust, the specific gravity (SG) of the must. It should be 1.105.

5. Check, and if necessary adjust, the temperature of the must. It should be 75°F (23°C).

6. Add the yeast to a cup of warm water. Let stand for 10 minutes. Stir in.

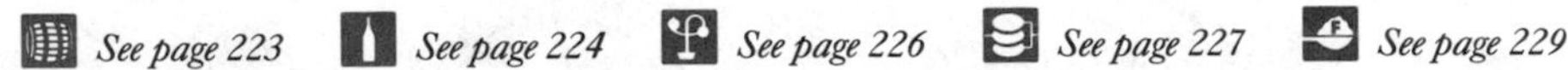
See page 220

See page 223

See page 224

See page 226

See page 227

See page 229

7. Cover the fermentor with a plastic sheet; tie down. Keep in a warm place (75°F [23°C]). After 24 hours, check that fermentation has begun. Foam should be visible on the surface and/or bubbles should be audible. If fermentation has not begun, see "Stuck Ferment," page 269.

8. Stir twice daily to keep the floating fruit moist.

9. Check SG every other day.

SECONDARY SEQUENCE

10. When SG reaches 1.020, scoop the berries into a straining bag and squeeze the juice gently into the fermentor. Discard the pulp.

11. Rack into a clean carboy. Top up with cold tap water.

12. Attach the fermentation lock.

13. Move to a cooler location, ideally 65°F (18°C).

14. After 10 days or at SG 1.000, whichever comes first, rack into a clean carboy. Top up with cold tap water.

15. After 3 weeks or at SG .990–.995, whichever comes first, rack into a clean carboy.

16. Add the finings. Top up with cold tap water. Let rest 10 days.

17. Rack into the primary fermentor.

18. Filter into a clean carboy.

19. Add ¼ teaspoon of sulphite crystals dissolved in a small amount of water. Add the Sinatin 17. Top up with cold tap water.

20. Bulk age 6 months.

21. Bottle.

22. Bottle age 6 months.

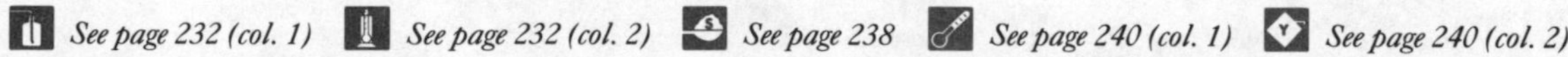

See page 232 (col. 1) *See page 232 (col. 2)* 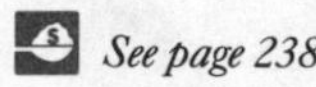 *See page 238*  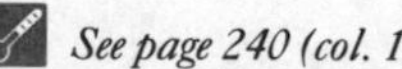 *See page 240 (col. 1)*  *See page 240 (col. 2)*

Saskatoon Berry Dessert Wine

READY: 1 YEAR

An interesting aperitif or dessert wine with 16% alcohol; best flavored with cinchona

PRIMARY INGREDIENTS

12 lb	Saskatoon berries (crushed)	5.5 kg
1 qt	Red grape concentrate	1 lit
12 lb	Sugar	5.5 kg
10 tsp	Vinacid O	
6 qt	HOT water	6 lit
2 tsp	Yeast nutrient	
2 tsp	Pectic enzyme	
1 tsp	Liquid tannin	
8	Campden tablets (crushed)	
8 qt	COLD water	8 lit
1 pkt	Wine yeast with a high alcohol tolerance	

SECONDARY INGREDIENTS

1½ cups	Sugar syrup (See "Syrup Feeding," page 274.)	360 ml

(continues)

	Gelatin finings	
1 oz	Sinatin 17	30 ml
	Cinchona (Follow manufacturer's instructions as to quantity.)	
¼ tsp	Sulphite crystals	
10 oz	Wine conditioner	300 ml

EQUIPMENT

Basic 10 + straining bag and potato masher; coarse filter pads

PRIMARY SEQUENCE

1. Crush the saskatoon berries and place them, together with the grape concentrate, in the primary fermentor. Add the hot water, sugar, and Vinacid O.

2. Stir thoroughly until all the sugar is dissolved.

3. Add the next 5 ingredients. Mix well.

4. Check, and if necessary adjust, the specific gravity (SG) of the must. It should be 1.100.

See page 220 See page 223 See page 224 See page 226 See page 227 See page 229

5. Check, and if necessary adjust, the temperature of the must. It should be 75°F (23°C).

6. Add the yeast to a cup of warm water. Let stand for 10 minutes. Stir in.

7. Cover the fermentor with a plastic sheet; tie down. Keep in a warm place (75°F [23°C]). After 24 hours, check that fermentation has begun. Foam should be visible on the surface and/or bubbles should be audible. If fermentation has not begun, see "Stuck Ferment," page 269.

8. Stir twice daily to keep the floating fruit moist.

9. Check the SG every other day.

SECONDARY SEQUENCE

10. When SG reaches 1.020, scoop the saskatoon berries into a straining bag and squeeze the juice out gently into the fermentor. Discard the pulp.

11. Rack into a clean carboy. Top up with cold tap water.

12. Attach the fermentation lock. Leave the carboy in a warm place.

13. After 10 days or at SG 1.000, whichever comes first, rack into a clean carboy.

14. When SG falls to .995, add 1½ cups of sugar syrup and leave the carboy in a warm place. When SG falls to .995 again, add a further 1½ cups of sugar syrup. Repeat until no further fermentation is possible.

15. When the fermentation stops, rack into a clean carboy. Add the cinchona and Sinatin 17. Top up with cold tap water.

16. Bulk age 3 months.

17. Add finings and let rest 10 days.

18. Filter into a clean carboy.

19. Add ¼ teaspoon of sulphite crystals dissolved in a small amount of water. Top up with tap water.

20. Bulk age 3 months.

21. Add sufficient wine conditioner to bring SG up to 1.100.

22. Bottle.

23. Bottle age 4 months.

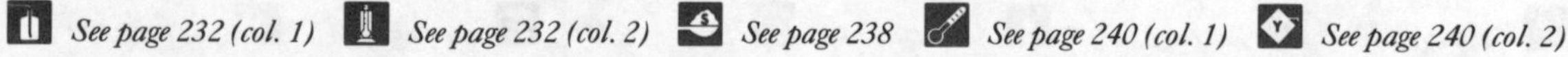
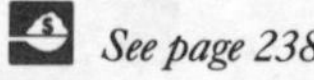

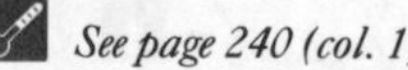

See page 232 (col. 1) *See page 232 (col. 2)* *See page 238* *See page 240 (col. 1)* *See page 240 (col. 2)*

Strawberry Red Rosé

READY: 7 MONTHS

A social wine much romanticized by country and western singers. Add some 7 Up for a delicious cooler.

PRIMARY INGREDIENTS

16 lb	Strawberries (crushed)	7.3 kg
1 qt	Red grape concentrate	1 lit
10 lb	Sugar	4.5 kg
6 qt	HOT water	6 lit
4 tsp	Vinacid R	
2 tsp	Liquid tannin	
2 tsp	Yeast nutrient	
2 tsp	Pectic enzyme	
8	Campden tablets (crushed)	
8 qt	COLD water	8 lit
1 pkt	Narbonne wine yeast	

SECONDARY INGREDIENTS

	Bentonite finings	
¼ tsp	Sulphite crystals	
10 oz	Wine conditioner	300 ml

EQUIPMENT

Basic 10 + straining bag and potato masher; fine filter pads

PRIMARY SEQUENCE

1. Crush the strawberries and place them, together with the grape concentrate, in the primary fermentor. Add the hot water, sugar, and Vinacid R.

2. Stir thoroughly until all the sugar is dissolved.

3. Add the next 5 ingredients. Mix well.

4. Check, and if necessary adjust, the temperature of the must. It should be 75°F (23°C).

5. Add the yeast to a cup of warm water. Let stand for 10 minutes. Stir in.

6. Cover the fermentor with a plastic sheet; tie down. Keep in a warm place (75°F [23°C]). After 24 hours, check that fermentation has begun. Foam should be visible on the surface and/or bubbles should be audible. If fermentation has not begun, see "Stuck Ferment," page 269.

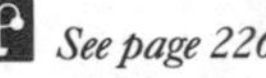
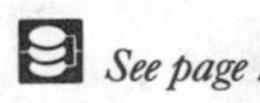
See page 220 *See page 223* *See page 224* *See page 226* *See page 227* *See page 229*

7. Stir twice daily to keep the floating fruit moist.

8. Check the specific gravity (SG) every other day.

SECONDARY SEQUENCE

9. When SG reaches 1.020, scoop the strawberries into a straining bag and squeeze the juice gently into the fermentor. Take care not to force the seeds through the mesh. Discard the pulp.

10. Rack into a clean carboy. Top up with cold tap water.

11. Attach the fermentation lock.

12. Move to a cooler location, ideally 65°F (18°C).

13. After 10 days or at SG 1.000, whichever comes first, rack into a clean carboy. Top up with cold tap water.

14. After 3 weeks or at SG .990–.995, whichever comes first, rack into a clean carboy.

15. Add the finings. Top up with cold tap water. Let rest 10 days.

16. Rack into the primary fermentor.

17. Filter into a clean carboy.

18. Add ¼ teaspoon of sulphite crystals in a small amount of water. Top up with cold tap water.

19. Bulk age 1 month.

20. Add the wine conditioner and bottle.

21. Bottle age 5 months.

See page 232 (col. 1)  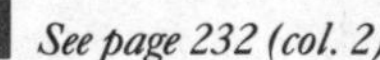 *See page 232 (col. 2)*   *See page 238* 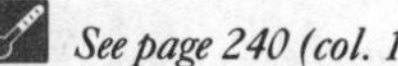 *See page 240 (col. 1)*  *See page 240 (col. 2)*

Cranberry Pleaser

READY: 9 MONTHS

A wonderful complement to turkey, this delicate wine can also be served as a cooler with 7 Up or alone over ice for sipping on a hot day.

PRIMARY INGREDIENTS

lb	Fresh or frozen cranberries (crushed)	6.8 kg
1 qt	Red grape concentrate	1 lit
10 lb	Sugar	4.5 kg
2 tsp	Vinacid R	
6 qt	HOT water	6 lit
2 tsp	Yeast nutrient	
2 tsp	Pectic enzyme	
1 tsp	Liquid tannin	
8	Campden tablets (crushed)	
8 qt	COLD water	8 lit
1 pkt	Narbonne wine yeast	

SECONDARY INGREDIENTS

	Gelatin finings	
¼ tsp	Sulphite crystals	
10 oz	Wine conditioner	300 ml

EQUIPMENT

Basic 10 + straining bag and potato masher; fine filter pads

PRIMARY SEQUENCE

1. Thaw the cranberries at room temperature in the primary fermentor. Crush them.

2. Add the grape concentrate, hot water, sugar, and Vinacid R.

3. Stir thoroughly until all the sugar is dissolved.

4. Add the next 5 ingredients. Mix well.

5. Check, and if necessary adjust, the specific gravity (SG) of the must. It should be 1.100.

6. Check, and if necessary adjust, the temperature of the must. It should be 75°F (23°C).

7. Add the yeast to a cup of warm water. Let stand for 10 minutes. Stir in.

8. Cover the fermentor with a plastic sheet; tie down. Keep in a warm place (75°F [23°C]). After 24 hours, check that fermentation has begun. Foam should be visible on the surface

See page 220
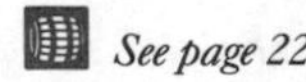
See page 223
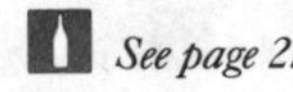
See page 224

See page 226

See page 227
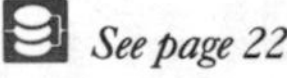
See page 229

and/or bubbles should be audible. If fermentation has not begun, see "Stuck Ferment," page 269.

9. Stir twice daily to keep the floating fruit moist.

10. Check the SG every other day.

SECONDARY SEQUENCE

11. When SG reaches 1.020, scoop the cranberries into a straining bag and squeeze the juice gently into the fermentor. Discard the pulp.

12. Rack into a clean carboy. Top up with cold tap water.

13. Attach the fermentation lock.

14. Move to a cooler location, ideally 65°F (18°C).

15. After 10 days or at SG 1.000, whichever comes first, rack into a clean carboy. Top up with cold tap water.

16. After 3 weeks or at SG .990–.995, whichever comes first, rack into a clean carboy.

17. Add the finings. Top up with cold tap water. Let rest 10 days.

18. Rack into the primary fermentor.

19. Filter into a clean carboy.

20. Add ¼ teaspoon of sulphite crystals dissolved in a small amount of water. Top up with cold tap water.

21. Bulk age 1 month.

22. Add the wine conditioner and bottle.

23. Bottle age 7 months.

See page 232 (col. 1)  *See page 232 (col. 2)*  *See page 238*  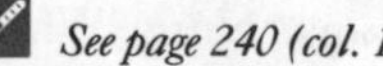 *See page 240 (col. 1)*  *See page 240 (col. 2)*

HARD FRUIT WINES

QUICK REFERENCE FOR HARD FRUIT WINES

Washing

If fruit has been sprayed recently with insect repellent or foliant, wash it before processing.

Fruit Quality

Do not use overripe, moldy, or fermenting fruit.

Fresh or Frozen?

You may use fresh or frozen fruit. Freezing improves fruit for winemaking because it breaks down cell structure and allows juice to be released more easily. To freeze fruit until a time convenient for winemaking, sprinkle it with sulphite solution (3 Campden tablets or ¼ tsp sulphite crystals per 2 cups water), then freeze it in 5 lb (2.2 kg) freezer bags. Note: When using fruit that has been sulphited before freezing, you do not need to add the sulphite listed in the primary recipe ingredients.

Crushing

Champagne cider requires apples to be crushed. For this you will need to rent or buy an apple crusher. (See "Crushing and Pressing Apples," page 248.)

When making scrumpy, or social wine from apples, crab apples, pears, and rhubarb, crushing is unnecessary; chopping the fruit is sufficient.

Pressing

Champagne cider requires apples to be pressed. For this you will need to rent or buy an apple press. (See "Crushing and Pressing Apples," page 248.)

When making scrumpy, or social wine from apples, crab apples, pears, and rhubarb, proceed as follows: Either place the fruit in a nylon straining bag and squeeze the juice out by hand, or scoop the chopped fruit out of the primary fermentor into a stainless-steel or plastic colander and press the juice from the fruit with a wooden spoon or mallet.

Used Pulp

Discard the pulp of hard fruits. It cannot be used for a second run.

Apple and Honey Wine

READY: 7 MONTHS

Traditionally, apples and honey fermented are called cyser. But in this recipe we have increased the alcohol content to 11% to make a wine. For a pleasant variation, add a small amount of vermouth flavoring.

PRIMARY INGREDIENTS

1 level teaspoon of sulphite crystals dissolved in 1 liter of water — to be sprinkled over the apples as they are chopped.

15 lb	Ripe apples (chopped)	6.8 kg
4 lb	Mild-flavored honey	1.8 kg
8 lb	Sugar	3.6 kg
6 tsp	Vinacid R	
6 qt	HOT water	6 lit
2 tsp	Pectic enzyme	
2 tsp	Yeast nutrient	
2 tsp	Liquid tannin	
8 qt	COLD water	8 lit
1 pkt	Champagne wine yeast	

SECONDARY INGREDIENTS

	Bentonite finings	
¼ tsp	Sulphite crystals	
8 oz	Wine conditioner	240 ml

EQUIPMENT

Basic 10 + straining bag; fine filter pads

PRIMARY SEQUENCE

1. Check that each apple is sound. Discard and replace unsound fruit.
2. Chop the apples into eighths and sprinkle the sulphite solution over the chopped pieces as you go, to keep them from browning. Do not use more than 2 cups of sulphite solution per 5-gallon batch of wine.
3. Place the chopped apples in the primary fermentor. Add the hot water, sugar, honey, and Vinacid R.
4. Stir thoroughly until all the sugar is dissolved.
5. Add the next 4 ingredients. Mix well.

 See page 220  See page 223  See page 224 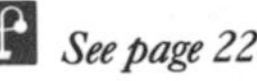 See page 226  See page 227  See page 229

6. Check, and if necessary adjust, the specific gravity (SG) of the must. It should be 1.095.

7. Check, and if necessary adjust, the temperature of the must. It should be 75°F (23°C).

8. Add the yeast to a cup of warm water. Let stand for 10 minutes. Stir in.

9. Cover the fermentor with a plastic sheet; tie down. Keep in a warm place (75°F [23°C]). After 24 hours, check that fermentation has begun. Foam should be visible on the surface and/or bubbles should be audible. If fermentation has not begun, see "Stuck Ferment," page 269.

10. Stir twice daily to keep the floating fruit moist.

11. Check SG every other day.

SECONDARY SEQUENCE

12. When SG reaches 1.020, scoop the apples into a straining bag and squeeze as dry as possible into the fermentor. Discard the pulp.

13. Rack into a clean carboy. Top up with cold tap water.

14. Attach the fermentation lock.

15. Move to a cooler location, ideally 65°F (18°C).

16. After 10 days or at SG 1.000, whichever comes first, rack into a clean carboy. Top up with cold tap water.

17. After 3 weeks or at SG .990–.995, whichever comes first, rack into a clean carboy.

18. Add the finings. Top up with cold tap water. Let rest 10 days.

19. Rack into the primary fermentor.

20. Filter into a clean carboy.

21. At this time, we suggest you test and adjust the acid content. The ideal acid reading is 6.5 g/lit. A lower acid reading will result in a flat-tasting wine.

22. Add ¼ teaspoon of sulphite crystals dissolved in a small amount of water. Top up with cold tap water.

23. Bulk age 1 month.

24. Add the wine conditioner and bottle.

25. Bottle age 5 months.

See page 232 (col. 1)  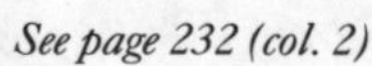 *See page 232 (col. 2)*  *See page 238*   *See page 240 (col. 1)*  *See page 240 (col. 2)*

Apple Wine

READY: 9 MONTHS

If sweetened, this light-bodied wine is similar to a California-style Chablis. Take extra care to avoid oxidation.

PRIMARY INGREDIENTS

1 level tsp sulphite crystals dissolved in 1 liter of water — to be sprinkled over the apples as they are chopped.

28 lb	Ripe apples (chopped)	12.7 kg
10 lb	Sugar	4.5 kg
2 tsp	Vinacid R	
6 qt	HOT water	6 lit
2 tsp	Yeast nutrient	
4 tsp	Liquid tannin	
4 tsp	Pectic enzyme	
10 qt	COLD water	10 lit
1 pkt	Champagne wine yeast	

SECONDARY INGREDIENTS

½ tsp	Sulphite crystals	
8 oz	Wine conditioner	240 ml
	Claro K. C. finings	

EQUIPMENT

Basic 10; fine filter pads

PRIMARY SEQUENCE

1. Check that each apple is sound. Discard and replace unsound fruit.

2. Chop the apples into eighths and sprinkle the sulphite solution on the chopped pieces as you go, to keep them from browning. Do not use more than 2 cups of sulphite solution per 5-gallon batch of wine.

3. Place the chopped apples in the primary fermentor. Add the hot water, sugar, and Vinacid R.

4. Stir thoroughly until all the sugar is dissolved.

5. Add the next 4 ingredients. Mix well.

6. Check, and if necessary adjust, the temperature of the must. It should be 75°F (23°C).

7. Add the yeast to a cup of warm water. Let stand for 10 minutes. Stir in.

 See page 220   See page 223  See page 224   See page 226  See page 227  See page 229

8. Cover the fermentor with a plastic sheet; tie down. Keep in a warm place (75°F [23°C]). After 24 hours, check that fermentation has begun. Foam should be visible on the surface and/or bubbles should be audible. If fermentation has not begun, see "Stuck Ferment," page 269.

9. Stir twice daily to keep the floating fruit moist.

10. Check the specific gravity (SG) every other day.

SECONDARY SEQUENCE

11. When SG reaches 1.020, rack into a clean carboy. Do not stir during the 8 hours prior to syphoning. Do not attempt to squeeze juice from the apple pulp; leave it undisturbed at the bottom of the fermentor and discard it when the racking is complete. Top up the carboy with cold tap water.

12. Attach the fermentation lock.

13. Move to a cooler location, ideally 65°F (18°C).

14. After 10 days or at SG 1.000, whichever comes first, rack into a clean carboy. Top up with cold tap water.

15. After 3 weeks, rack into a clean carboy.

16. Add ¼ teaspoon of sulphite crystals dissolved in a small amount of water and top up with cold tap water. Let rest 30 days.

17. At SG .990–.995, add the finings. Let rest 10 days.

18. Rack into the primary fermentor.

19. Filter into a clean carboy.

20. Add the wine conditioner and ¼ teaspoon of sulphite crystals dissolved in a small amount of water.

21. Bulk age 4 months.

22. Bottle.

23. Bottle age 4 months.

See page 232 (col. 1)   *See page 232 (col. 2)*  *See page 238*   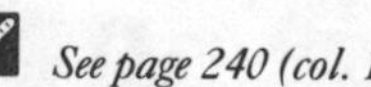 *See page 240 (col. 1)*  *See page 240 (col. 2)*

Scrumpy

READY: 4 MONTHS

A rough, coarse cider drunk in England; often cloudy and quite strong. You can fine it and filter it, but it will still be darker in color than fine cider.

PRIMARY INGREDIENTS

1 level tsp of sulphite crystals dissolved in 2 cups of water — to be sprinkled over the apples as they are chopped.

120 lb	Ripe apples (chopped)	55 kg
2 tsp	Pectic enzyme	
2 tsp	Yeast nutrient	
2 tsp	Liquid tannin	
1 pkt	Bayanus wine yeast	

SECONDARY INGREDIENTS

	Claro K. C. finings	
8 oz	Cane sugar	240 g
1 pkt	Champagne yeast	

EQUIPMENT

Basic 10 + straining bag

PRIMARY SEQUENCE

1. Check that each apple is sound; discard and replace unsound fruit.

2. Chop the apples into eighths and sprinkle the sulphite solution over the chopped pieces as you go, to keep them from browning.

3. Place the chopped apples in the primary fermentor. Add the next 3 ingredients.

4. Stir thoroughly.

5. Check, and if necessary adjust, the temperature of the must. It should be 75°F (23°C).

6. Add the yeast to a cup of warm water. Let stand for 10 minutes. Stir in.

7. Cover the fermentor with a plastic sheet; tie down. Keep in a warm place (75°F [23°C]). After 24 hours, check that fermentation has begun. Foam should be visible on the surface and/or bubbles should be audible. If fermentation has not begun, see "Stuck Ferment," page 269.

8. Stir twice daily to keep the floating fruit moist.

 See page 220  See page 223  See page 224    See page 226  See page 227  See page 229

SECONDARY SEQUENCE

9. After 5 days, rack into a clean carboy. Do not stir for 8 hours prior to syphoning. Place the syphon tip under the floating cap but above the sediment.

10. Place the pulp in a fine mesh bag and extract as much juice as possible. Add the juice to the carboy. Top up with cold tap water.

11. Attach the fermentation lock.

12. Move to a cooler location, ideally 65°F (18°C).

13. Check SG every other day.

14. At SG 1.000, rack into a clean carboy. Top up with cold tap water.

15. At SG .997 (after approximately 2 weeks), rack into a clean carboy.

16. Add the finings.

17. Add ¼ teaspoon of sulphite crystals dissolved in a small amount of water. Let rest 10 days.

18. Rack into a clean carboy. Add the sugar and the champagne yeast dissolved in a little wine.

19. Replace the fermentation lock. Top up with clean tap water.

20. After 12 hours, bottle in pressure-tolerant bottles (beer bottles, soda bottles, or champagne bottles).

21. Bottle age 3 months.

See page 232 (col. 1)  *See page 232 (col. 2)*  *See page 238* *See page 240 (col. 1)*  *See page 240 (col. 2)*

Champagne Cider

READY: 6–9 MONTHS

While this is a labor-intensive recipe, the resulting cider is very much like champagne — even the nose is similar. Your only complaint will be that not enough was made.

PRIMARY INGREDIENTS

1 level tsp of sulphite crystals dissolved in 1 liter of water — to be sprinkled over the apples as they are crushed.

125 lb	Apples	57 kg
2 tsp	Pectic enzyme	
2 tsp	Yeast nutrient	
2 tsp	Liquid tannin	
2 pkt	Champagne yeast	

SECONDARY INGREDIENTS

	Claro K. C. finings	
10 oz	Wine conditioner	300 ml

EQUIPMENT

Basic 10 + apple crusher, apple press, straining bag, acid-testing kit; fine filter pads

PRIMARY SEQUENCE

1. Crush the apples. (See "Crushing and Pressing Apples," page 248.)

2. Press the apples. Run the juice into a pail, then pour it into the primary fermentor.

3. Check, and if necessary adjust, the specific gravity (SG) of the must. It should be 1.060.

4. Check, and if necessary adjust, the temperature of the must. It should be 70°F (21°C).

5. Check, and if necessary adjust, the acid level. It should be 4 g/lit, but 5 g/lit is tolerable.

6. Allow the apple juice to stand 8 hours. A heavy sediment will settle. Rack into a clean primary fermentor.

7. Add the yeast to a cup of warm water. Let stand for 10 minutes. Stir in.

8. Add the rest of the primary ingredients.

9. Cover the fermentor with a plastic sheet; tie down. Keep in a warm place (70°F [21°C]). After 24 hours, check that fermentation has begun. Foam should be visible on the surface and/or bubbles should be audible. If

See page 220

See page 223
See page 224

See page 226

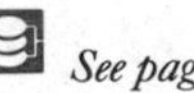
See page 227

See page 229

fermentation has not begun, see "Stuck Ferment," page 269.

10. Check SG every day.

SECONDARY SEQUENCE

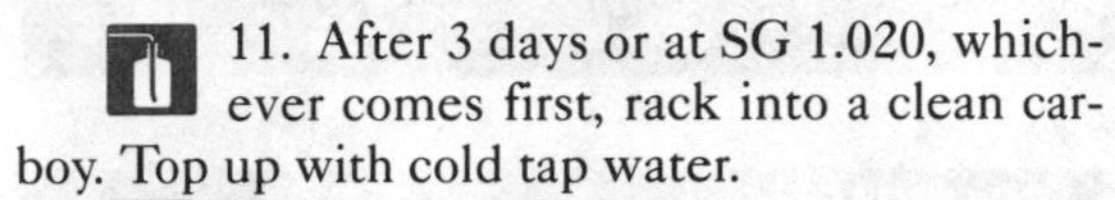

11. After 3 days or at SG 1.020, whichever comes first, rack into a clean carboy. Top up with cold tap water.

12. Attach the fermentation lock.

13. Move to a cooler location, ideally 65°F (18°C).

14. Check the SG every other day.

15. At SG 1.000 (after approximately 7 days), rack into a clean carboy. Top up with cold tap water.

16. At SG .997 (after approximately 2 weeks), rack into a clean carboy.

17. Bulk age 1 month.

18. Add finings and let rest 10 days.

19. Filter into a clean carboy. Sweeten to taste with wine conditioner; use a minimum of 2 ounces (60 ml) per gallon.

20. Bottle.

21. Bottle age 3–6 months.

22. Sparkle with CO_2. (See "Champagne — Dispatch Method," page 214.)

See page 232 (col. 1)  *See page 232 (col. 2)*  *See page 238*  *See page 240 (col. 1)*  *See page 240 (col. 2)*

Crab Apple Social Wine

READY: 1 YEAR

Often becomes a blush wine; a delicious use of crab apples from your garden

PRIMARY INGREDIENTS

1 level tsp sulphite crystals dissolved in 1 liter of water — to be sprinkled over the apples as they are chopped.

16 lb	Ripe crab apples (finely chopped)	7.3 kg
9 lb	Sugar	4 kg
1 qt	White grape concentrate	1 lit
3 tsp	Vinacid O	
6 qt	HOT water	6 lit
2 tsp	Yeast nutrient	
2 tsp	Pectic enzyme	
8 qt	COLD water	8 lit
1 pkt	Champagne wine yeast	

SECONDARY INGREDIENTS

	Bentonite finings	
¼ tsp	Sulphite crystals	
10 oz	Wine conditioner	300 ml

EQUIPMENT

Basic 10 + straining bag; fine filter pads

PRIMARY SEQUENCE

1. Check that each apple is sound. Discard and replace unsound fruit.

2. Chop the crab apples finely, sprinkling the sulphite solution on the chopped pieces as you go, to keep them from browning. Do not use more than 2 cups of sulphite solution per 5-gallon batch of wine.

3. Place the chopped crab apples and the grape concentrate in the primary fermentor. Add the hot water, sugar, and Vinacid O.

4. Stir thoroughly until all the sugar is dissolved.

5. Add the next 3 ingredients. Mix well.

6. Check, and if necessary adjust, the specific gravity (SG) of the must. It should be 1.100.

7. Check, and if necessary adjust, the temperature of the must. It should be 75°F (23°C).

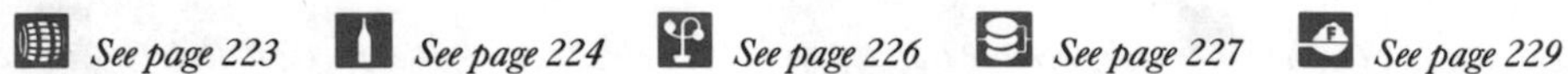

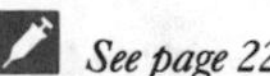
See page 220

See page 223
See page 224

See page 226

See page 227

See page 229

8. Add the yeast to a cup of warm water. Let stand for 10 minutes. Stir in.

9. Cover the fermentor with a plastic sheet; tie down. Keep in a warm place (75°F [23°C]). After 24 hours, check that fermentation has begun. Foam should be visible on the surface and/or bubbles should be audible. If fermentation has not begun, see "Stuck Ferment," page 269.

10. Stir twice daily to keep the floating fruit moist.

11. Check SG every other day.

SECONDARY SEQUENCE

12. When SG reaches 1.020, scoop the apples into a straining bag and squeeze the juice gently into the fermentor. Discard the pulp.

13. Rack into a clean carboy. Top up with cold tap water.

14. Attach the fermentation lock.

15. Move to a cooler location, ideally 65°F (18°C).

16. After 10 days or at SG 1.000, whichever comes first, rack into a clean carboy. Top up with cold tap water.

17. After 3 weeks or at SG .990–.995, whichever comes first, rack into a clean carboy.

18. Add the finings. Top up with cold tap water. Let rest 10 days.

19. Rack into the primary fermentor.

20. Filter into a clean carboy.

21. Add ¼ teaspoon of sulphite crystals dissolved in a small amount of water. Top up with cold tap water.

22. Bulk age 3 months.

23. Add the wine conditioner and bottle.

24. Bottle age 9 months.

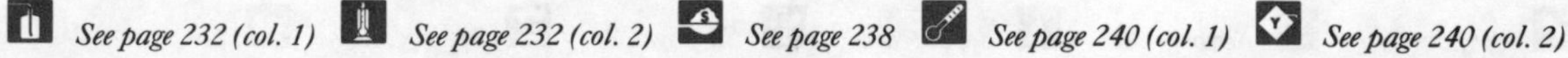
See page 232 (col. 1) *See page 232 (col. 2)*  *See page 238*  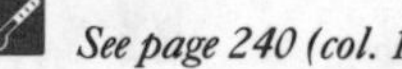 *See page 240 (col. 1)*  *See page 240 (col. 2)*

Pear Wine

READY: 7 MONTHS

A light refreshing wine from pears, without the expense of crushing and pressing equipment

PRIMARY INGREDIENTS

1 level tsp sulphite crystals dissolved in 1 liter of water — to be sprinkled over the pears as they are chopped.

28 lb	Ripe pears (chopped)	12.7 kg
10 lb	Sugar	4.5 kg
2 tsp	Vinacid R	
6 qt	HOT water	6 lit
2 tsp	Yeast nutrient	
2 tsp	Pectic enzyme	
2 tsp	Liquid tannin	
9 qt	COLD water	9 lit
1 pkt	Champagne wine yeast	

SECONDARY INGREDIENTS

	Bentonite finings	
½ tsp	Sulphite crystals	
8 oz	Wine conditioner	240 ml

EQUIPMENT

Basic 10 + straining bag; fine filter pads

PRIMARY SEQUENCE

1. Check that each pear is sound. Discard and replace unsound fruit.

2. Chop the pears into eighths, sprinkling the sulphite solution on the chopped pieces as you go, to keep them from browning. Do not use more than 2 cups of solution per 5-gallon batch of wine.

3. Place the chopped pears in the primary fermentor. Add the hot water, sugar, and Vinacid R.

4. Stir thoroughly until all the sugar is dissolved.

5. Add the next 4 ingredients. Mix well.

6. Check, and if necessary adjust, the specific gravity (SG) of the must. It should be 1.100.

7. Check, and if necessary adjust, the temperature of the must. It should be 75°F (23°C).

See page 220  *See page 223*  *See page 224*  *See page 226*  *See page 227*  *See page 229*

8. Add the yeast to a cup of warm water. Let stand for 10 minutes. Stir in.

9. Cover the fermentor with a plastic sheet; tie down. Keep in a warm place (75°F [23°C]). After 24 hours, check that fermentation has begun. Foam should be visible on the surface and/or bubbles should be audible. If fermentation has not begun, see "Stuck Ferment," page 269.

10. Stir twice daily to keep the floating fruit moist.

11. Check SG every other day.

SECONDARY SEQUENCE

12. When SG reaches 1.020, scoop the pears into a straining bag and squeeze it as dry as possible into the fermentor. Discard the pulp.

13. Rack into a clean carboy. Top up with cold tap water.

14. Attach the fermentation lock.

15. Move to a cooler location, ideally 65°F (18°C).

16. After 10 days or at SG 1.000, whichever comes first, rack into a clean carboy. Top up with cold tap water.

17. After 3 weeks or at SG .990–.995, whichever comes first, rack into a clean carboy.

18. Add the finings. Top up with cold tap water. Let rest 10 days.

19. Rack into the primary fermentor.

20. Filter into a clean carboy.

21. Add ½ teaspoon of sulphite crystals dissolved in a small amount of water. Top up with cold tap water. (Note: Extra sulphite [½ tsp] is essential for pears.)

22. Bulk age 1 month.

23. Add the wine conditioner and bottle.

24. Bottle age 5 months.

 *See page 232 (col. 1)*  *See page 232 (col. 2)*  *See page 238* 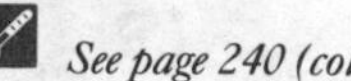 *See page 240 (col. 1)*  *See page 240 (col. 2)*

Rhubarb Wine

READY: 1 YEAR

A delightful white wine, with a fruity aroma

PRIMARY INGREDIENTS

12 lb	Rhubarb	5.5 kg
1 qt	White grape concentrate	1 lit
10 lb	Sugar	4.5 kg
6 qt	HOT water	6 lit
2 tsp	Yeast nutrient	
4 tsp	Liquid tannin	
2 tsp	Pectic enzyme	
8	Campden tablets (crushed)	
8 qt	COLD water	8 lit
1 pkt	Champagne wine yeast	

SECONDARY INGREDIENTS

	Claro K. C. finings	
¼ tsp	Sulphite crystals	
8 oz	Wine conditioner	240 ml

EQUIPMENT

Basic 10 + straining bag and potato masher; fine filter pads

PRIMARY SEQUENCE

1. Place the chopped rhubarb in the primary fermentor. Pour the sugar over the rhubarb and stir the two together. Cover with a plastic sheet for 24 hours.

2. Crush the rhubarb with a potato masher or a wooden mallet.

3. Pour the hot water over the crushed rhubarb and stir vigorously. When the rhubarb and water have been stirred, scoop the rhubarb into a straining bag and squeeze it as dry as possible into the fermentor. Discard the pulp.

4. Add the grape concentrate. Mix well.

5. Add the next 5 ingredients. Mix well.

6. Check, and if necessary adjust, the specific gravity (SG) of the must. It should be 1.110.

7. Check, and if necessary adjust, the temperature of the must. It should be 75°F (23°C).

8. Add the yeast to a cup of warm water. Let stand for 10 minutes. Stir in.

9. Cover the fermentor with a plastic sheet; tie down. Keep in a warm place (75°F [23°C]). After 24 hours, check that fermentation has

 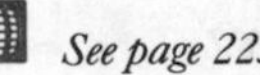 See page 220  See page 223  See page 224  See page 226 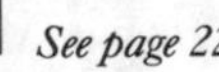  See page 227  See page 229

begun. Foam should be visible on the surface and/or bubbles should be audible. If fermentation has not begun, see "Stuck Ferment," page 269.

10. Check SG every other day.

SECONDARY SEQUENCE

11. When SG reaches 1.020, rack into a clean carboy. Top up with cold tap water.

12. Attach the fermentation lock.

13. Move to a cooler location, ideally 65°F (18°C).

14. After 10 days or at SG 1.000, whichever comes first, rack into a clean carboy. Top up with cold tap water.

15. After 3 weeks or at SG .990–.995, whichever comes first, rack into a clean carboy.

16. Add the finings. Top up with cold tap water. Let rest 10 days.

17. Rack into the primary fermentor.

18. Filter into a clean carboy.

19. Add ¼ teaspoon of sulphite crystals dissolved in a small amount of water. Top up with cold tap water.

20. Bulk age 6 months.

21. Add the wine conditioner and bottle.

22. Bottle age 6 months.

See page 232 (col. 1) *See page 232 (col. 2)*
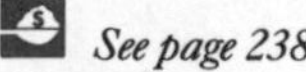
See page 238
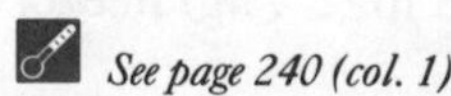
See page 240 (col. 1)

See page 240 (col. 2)

SOFT FRUIT WINES

QUICK REFERENCE FOR SOFT FRUIT WINES

Washing

If fruit has been sprayed recently with insect repellent or foliant, wash it before processing.

Fruit Quality

Do not use overripe, moldy, or fermenting fruit.

Fresh or Frozen?

You may use fresh or frozen fruit. Freezing improves fruit for winemaking because it breaks down cell structure and allows juice to be released more easily. To freeze fruit until a time convenient for winemaking, sprinkle it first with a sulphite solution (3 Campden tablets or ¼ teaspoon of sulphite crystals per 2 cups of water), then freeze in 5 lb (2.2 kg) freezer bags.

Note: When using fruit that has been sulphited before freezing, you do not need to add the sulphite listed in the primary recipe ingredients.

Crushing

Cherries and chokecherries need to be crushed. Remove the leaves and stems and crush with a potato masher. Take care not to crush the pits, or they will release a bitter taste into the must.

With apricots, peaches, or plums, crushing is unnecessary; simply halve the fruit and remove the pits.

Pressing

Cherries and chokecherries need to be pressed. Place them in a nylon straining bag and squeeze the juice out by hand, or scoop them into a stainless-steel or plastic colander and press the juice out with a mallet or a wooden spoon.

Apricots, peaches, and plums break down into a very soft pulp that can easily be forced through the mesh of the straining bag. When pressing these fruits, therefore, squeeze very gently and let gravity do most of the work. A heavy sediment should be avoided.

Used Pulp

Discard the pulp of soft fruits. It cannot be used for a second run.

Apricot Delight

READY: 7 MONTHS

Apricot is the most popular fruit base among winemakers. This recipe produces a pleasant, naturally sweetened social wine.

PRIMARY INGREDIENTS

96 oz	Apricot fruit wine base	3 lit
6 oz	Doradillo grape concentrate*	180 ml
9½ lb	Sugar	4.3 kg
8 tsp	Vinacid R	
6 qt	HOT water	6 lit
2 tsp	Yeast nutrient	
2 tsp	Pectic enzyme	
1 tsp	Liquid tannin	
5	Campden tablets (crushed)	
7 qt	COLD water	7 lit
1 pkt	Champagne wine yeast	

SECONDARY INGREDIENTS

	Bentonite finings	
¼ tsp	Sulphite crystals	
26 oz	Doradillo grape concentrate*	780 ml
2 tsp	Stabilizer	

(continues)

EQUIPMENT

Basic 10 + straining bag; fine filter pads

PRIMARY SEQUENCE

1. Pour the apricot fruit base and the grape concentrate into the primary fermentor. Add the hot water, sugar, and Vinacid R.

2. Stir thoroughly until all the sugar is dissolved.

3. Add the next 5 ingredients. Mix well.

4. Check, and if necessary adjust, the specific gravity (SG) of the must. It should be 1.090.

*Purchase 1 quart (1 lit) of Doradillo grape concentrate; use 6 ounces (180 ml) in the primary ingredients and refrigerate the remainder to use as a natural sweetener in the secondary ingredients.

See page 220  See page 223  See page 224   See page 226  See page 227  See page 229

5. Check, and if necessary adjust, the temperature of the must. It should be 75°F (23°C).

6. Add the yeast to a cup of warm water. Let stand for 10 minutes. Stir in.

7. Cover the fermentor with a plastic sheet; tie down. Keep in a warm place (75°F [23°C]). After 24 hours, check that fermentation has begun. Foam should be visible on the surface and/or bubbles should be audible. If fermentation has not begun, see "Stuck Ferment," page 269.

8. Stir twice daily to keep the fruit moist.

9. Check SG every other day.

SECONDARY SEQUENCE

10. When SG reaches 1.020, scoop the apricots into a straining bag and squeeze the juice very gently into the fermentor so that no pulp is forced through the mesh. Discard the pulp.

11. Rack into a clean carboy. Top up with cold tap water.

12. Attach the fermentation lock.

13. Move to a cooler location, ideally 65°F (18°C).

14. After 10 days or at SG 1.000, whichever comes first, rack into a clean carboy. Top up with cold tap water.

15. After 3 weeks or at SG .990–.995, whichever comes first, rack into a clean carboy.

16. Add the finings. Top up with cold tap water. Let rest 10 days.

17. Rack into the primary fermentor. Add the remainder of the grape concentrate.

18. Filter into a clean carboy.

19. Add ¼ teaspoon of sulphite crystals dissolved in a small amount of water. Add the stabilizer and stir well to mix. Top up with cold tap water.

20. Bulk age 1 month.

21. Bottle.

22. Bottle age 5 months.

See page 232 (col. 1)  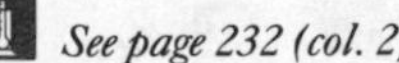 *See page 232 (col. 2)*  *See page 238*  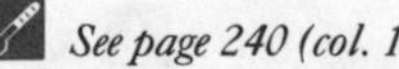 *See page 240 (col. 1)*  *See page 240 (col. 2)*

Apricot Wine

READY: 7 MONTHS

This recipe makes a medium-dry table wine. For a social wine, add an extra 1½ pounds (680 g) of sugar. For a German-style table wine, use 1 pound (450 g) less sugar.

PRIMARY INGREDIENTS

15 lb	Fresh ripe apricots	6.8 kg
1 qt	White grape concentrate	1 lit
9 lb	Sugar	4 kg
6 qt	HOT water	6 lit
2 tsp	Yeast nutrient	
4 tsp	Liquid tannin	
2 tsp	Pectic enzyme	
8	Campden tablets (crushed)	
8 qt	COLD water	8 lit
1 pkt	Champagne wine yeast	

SECONDARY INGREDIENTS

	Claro K. C. finings	
¼ tsp	Sulphite crystals	
8 oz	Wine conditioner	240 ml

EQUIPMENT

Basic 10 + straining bag; fine filter pads

PRIMARY SEQUENCE

1. Halve the apricots and remove their pits.

2. Place the apricots and the grape concentrate in the primary fermentor. Add the hot water and sugar.

3. Stir thoroughly until all the sugar is dissolved.

4. Add the next 5 ingredients. Mix well.

5. Check, and if necessary adjust, the specific gravity (SG) of the must. It should be 1.100.

6. Check, and if necessary adjust, the temperature of the must. It should be 75°F (23°C).

7. Add the yeast to a cup of warm water. Let stand for 10 minutes. Stir in.

8. Cover the fermentor with a plastic sheet; tie down. Keep in a warm place (75°F [23°C]). After 24 hours, check that fermentation has begun. Foam should be visible on the surface and/or bubbles should be audible. If fermenta-

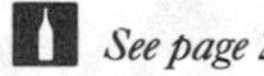

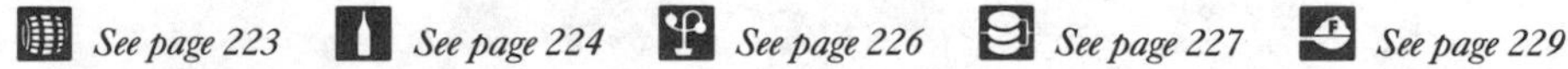
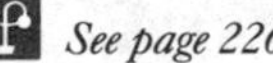

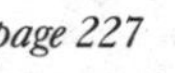
See page 220 See page 223 See page 224 See page 226 See page 227 See page 229

tion has not begun, see "Stuck Ferment," page 269.

9. Stir twice daily to keep the fruit moist.

10. Check SG every other day.

SECONDARY SEQUENCE

11. When SG reaches 1.020, scoop the apricots into a straining bag and squeeze the juice very gently into the fermentor so that no pulp is forced through the mesh. Discard the pulp.

12. Rack into a clean carboy. Top up with cold tap water.

13. Attach the fermentation lock.

14. Move to a cooler location, ideally 65°F (18°C).

15. After 10 days or at SG 1.000, whichever comes first, rack into a clean carboy. Top up with cold tap water.

16. After 3 weeks or at SG .990–.995, whichever comes first, rack into a clean carboy.

17. Add the finings. Top up with cold tap water. Let rest 10 days.

18. Rack into the primary fermentor.

19. Filter into a clean carboy.

20. Add ¼ teaspoon sulphite crystals dissolved in a small amount of water. Top up with cold tap water.

21. Bulk age 3 months.

22. Add the wine conditioner and bottle.

23. Bottle age 3 months.

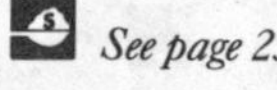

See page 232 (col. 1) *See page 232 (col. 2)* *See page 238* *See page 240 (col. 1)* *See page 240 (col. 2)*

Apricot Dessert Wine

READY: 1 YEAR

A delicious base for spritzers or coolers. Serve with sweet desserts or fruit.

PRIMARY INGREDIENTS

15 lb	Fresh ripe apricots	6.8 kg
1 qt	White grape concentrate	1 lit
12 lb	Sugar	5.5 kg
6 qt	HOT water	6 lit
2 tsp	Yeast nutrient	
4 tsp	Liquid tannin	
2 tsp	Pectic enzyme	
8	Campden tablets (crushed)	
8 qt	COLD water	8 lit
1 pkt	Wine yeast with a high alcohol tolerance	

SECONDARY INGREDIENTS

	Gelatin finings	
¼ tsp	Sulphite crystals	
12 oz	Wine conditioner	360 ml

EQUIPMENT

Basic 10 + straining bag; fine filter pads

PRIMARY SEQUENCE

1. Halve the apricots and remove their pits.

2. Place the apricots and the grape concentrate in the primary fermentor. Add the hot water and sugar.

3. Stir thoroughly until all the sugar is dissolved.

4. Add the next 5 ingredients. Mix well.

5. Add the yeast to a cup of warm water. Let stand for 10 minutes. Stir in.

6. Cover the fermentor with a plastic sheet; tie down. Keep in a warm place (75°F [23°C]). After 24 hours, check that fermentation has begun. Foam should be visible on the surface and/or bubbles should be audible. If fermentation has not begun, see "Stuck Ferment," page 269.

7. Stir twice daily to keep the floating fruit moist.

8. Check the specific gravity (SG) every other day.

 *See page 220*  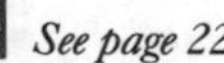 *See page 223*  *See page 224*  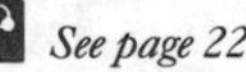 *See page 226*  *See page 227*  *See page 229*

SECONDARY SEQUENCE

9. When SG reaches 1.020, scoop the apricots into a straining bag and squeeze the juice very gently into the fermentor so that no pulp is forced through the mesh. Discard the pulp.

10. Rack into a clean carboy. Top up with cold tap water.

11. Attach the fermentation lock.

12. Move to a cooler location, ideally 65°F (18°C).

13. After 10 days or at SG 1.000, whichever comes first, rack into a clean carboy. Top up with cold tap water.

14. After 3 weeks, rack into a clean carboy. Top up with cold tap water.

15. When the wine has stopped fermenting, add the finings and let rest 10 days.

16. Rack into a clean carboy.

17. Bulk age 1 month.

18. Filter into a clean carboy.

19. Add ¼ teaspoon of sulphite crystals dissolved in a small amount of water. Top up with cold tap water.

20. Check SG. If it is lower than 1.000, make it up to that level with wine conditioner.

21. Bottle.

22. Bottle age 10 months.

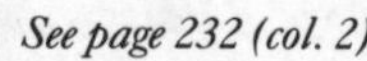

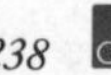
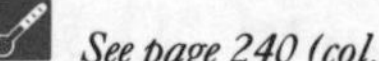

See page 232 (col. 1) *See page 232 (col. 2)* *See page 238* *See page 240 (col. 1)* *See page 240 (col. 2)*

Cherry Red Table Wine

READY: 13 MONTHS

If you can harvest your cherries before the birds do, they make a very good table wine.

PRIMARY INGREDIENTS

15 lb	Sweet red cherries (crushed)	6.8 kg
1 qt	Red grape concentrate	1 lit
13 lb	Sugar	6 kg
4 tsp	Vinacid O	
6 qt	HOT water	6 lit
2 tsp	Yeast nutrient	
2 tsp	Pectic enzyme	
2 tsp	Liquid tannin	
8	Campden tablets (crushed)	
8 qt	COLD water	8 lit
1 pkt	Montpellier wine yeast	

SECONDARY INGREDIENTS

	Bentonite finings	
¼ tsp	Sulphite crystals	
1 oz	Sinatin 17	30 ml

EQUIPMENT

Basic 10 + straining bag; coarse filter pads

PRIMARY SEQUENCE

1. Crush the cherries and place them, together with the grape concentrate, in the primary fermentor. Add the hot water, sugar, and Vinacid O.

2. Stir thoroughly until all the sugar is dissolved.

3. Add the next 5 ingredients. Mix well.

4. Check, and if necessary adjust, the specific gravity (SG) of the must. It should be 1.010.

5. Check, and if necessary adjust, the temperature of the must. It should be 75°F (23°C).

6. Add the yeast to a cup of warm water. Let stand for 10 minutes. Stir in.

7. Cover the fermentor with a plastic sheet; tie down. Keep in a warm place (75°F [23°C]). After 24 hours, check that fermentation has begun. Foam should be visible on the surface and/or bubbles should be audible. If fermenta-

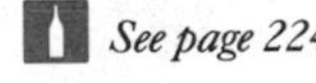

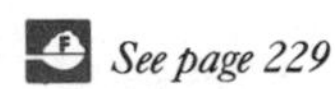
See page 220 See page 223 See page 224 See page 226 See page 227 See page 229

tion has not begun, see "Stuck Ferment," page 269.

8. Stir twice daily to keep the floating fruit moist.

9. Check SG every other day.

SECONDARY SEQUENCE

10. When SG reaches 1.020, scoop the cherries into a straining bag and squeeze as dry as possible into the fermentor. Discard the pulp.

11. Rack into a clean carboy. Top up with cold tap water.

12. Attach fermentation lock.

13. Move to a cooler location, ideally 65°F (18°C).

14. After 10 days or at SG 1.000, whichever comes first, rack into a clean carboy. Top up with cold tap water.

15. After 3 weeks or at SG .990–.995, whichever comes first, rack into a clean carboy.

16. Add the finings. Top up with cold tap water.

17. Rack into the primary fermentor.

18. Filter into a clean carboy.

19. Add ¼ teaspoon of sulphite crystals dissolved in a small amount of water. Add the Sinatin 17. Top up with cold tap water.

20. Bulk age 3 months.

21. Bottle.

22. Bottle age 9 months.

See page 232 (col. 1)  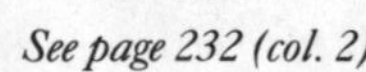 *See page 232 (col. 2)*   *See page 238*  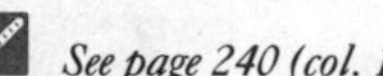 *See page 240 (col. 1)*  *See page 240 (col. 2)*

Cherry Port

READY: 18 MONTHS

Because cherries ferment well, they lend themselves to high-alcohol wines. If this wine is aged 3 years, it will become a tawny port.

PRIMARY INGREDIENTS

32 lb	Sweet red cherries (crushed)	14.5 kg
12 lb	Sugar	5.5 kg
2 tsp	Vinacid O	
6 qt	HOT water	6 lit
2 tsp	Yeast nutrient	
2 tsp	Pectic enzyme	
2 tsp	Liquid tannin	
8	Campden tablets (crushed)	
7 qt	COLD water	7 lit
1 pkt	Wine yeast with a high alcohol tolerance	

SECONDARY INGREDIENTS

3 cups	Sugar syrup (See "Syrup Feeding," page 274.)	

(continues)

	Gelatin finings	
¼ tsp	Sulphite crystals	
1½ oz	Sinatin 17	45 ml
26 oz	Vodka	780 ml
12 oz	Wine conditioner	360 ml

EQUIPMENT

Basic 10 + straining bag and potato masher; coarse filter pads

PRIMARY SEQUENCE

1. Crush the cherries and place them in the primary fermentor. Add the hot water, sugar, and Vinacid O.

2. Stir thoroughly until all the sugar is dissolved.

3. Add the next 5 ingredients. Mix well.

4. Check, and if necessary adjust, the specific gravity (SG) of the must. It should be 1.110.

5. Check, and if necessary adjust, the temperature of the must. It should be 75°F (23°C).

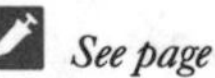

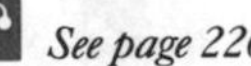

See page 220 See page 223 See page 224 See page 226 See page 227 See page 229

6. Add the yeast to a cup of warm water. Let stand for 10 minutes. Stir in.

7. Cover the fermentor with a plastic sheet; tie down. Keep in a warm place (75°F [23°C]). After 24 hours, check that fermentation has begun. Foam should be visible on the surface and/or bubbles should be audible. If fermentation has not begun, see "Stuck Ferment," page 269.

8. Stir twice daily to keep the floating fruit moist.

9. Check SG every other day.

SECONDARY SEQUENCE

10. When SG reaches 1.020, scoop the cherries into a straining bag and squeeze as dry as possible into the fermentor. Discard the pulp.

11. Rack into a clean carboy. Top up with cold tap water.

12. Attach the fermentation lock and leave the fermentor in its warm location.

13. When SG falls to 1.000 or below, add 2 cups of cooled sugar syrup. Continue to ferment in a warm place.

14. If SG falls below 1.000 again, add the remainder of the sugar syrup. If the SG does not fall below 1.000, do not add the remainder of the syrup. Continue to ferment in a warm place.

15. When the fermentation ceases, rack into a clean carboy. Top up with cold tap water. Let rest 1 month.

16. Add the finings. Let rest 10 days.

17. Rack into the primary fermentor.

18. Filter into a clean carboy.

19. Add ¼ teaspoon of sulphite crystals dissolved in a small amount of water. Add the Sinatin 17. Top up with cold tap water.

20. Bulk age 3 months.

21. Rack into the primary fermentor. Add the vodka and the wine conditioner.

22. Bottle.

23. Bottle age 1 year.

See page 232 (col. 1)  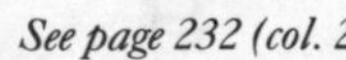 *See page 232 (col. 2)*  *See page 238*   *See page 240 (col. 1)*  *See page 240 (col. 2)*

Cherry Rosé

READY: 7 MONTHS

A slightly sweet, very pleasant social wine

PRIMARY INGREDIENTS

15 lb	Sweet red cherries (crushed)	6.8 kg
1 qt	White grape concentrate	1 lit
10 lb	Sugar	4.5 kg
7 tsp	Vinacid R	
6 qt	HOT water	6 lit
2 tsp	Yeast nutrient	
2 tsp	Pectic enzyme	
2 tsp	Liquid tannin	
8	Campden tablets (crushed)	
8 qt	COLD water	8 lit
1 pkt	Narbonne wine yeast	

SECONDARY INGREDIENTS

	Claro K. C. finings	
¼ tsp	Sulphite crystals	
8 oz	Wine conditioner	240 ml

EQUIPMENT

Basic 10 + straining bag and potato masher; fine filter pads

PRIMARY SEQUENCE

1. Crush the cherries and place them, together with the grape concentrate, in the primary fermentor. Add the hot water, sugar, and Vinacid R.

2. Stir thoroughly until all the sugar is dissolved.

3. Add the next 5 ingredients. Mix well.

4. Check, and if necessary adjust, the specific gravity (SG) of the must. It should be 1.095.

5. Check, and if necessary adjust, the temperature of the must. It should be 75°F (23°C).

6. Add the yeast to a cup of warm water. Let stand for 10 minutes. Stir in.

7. Cover the fermentor with a plastic sheet; tie down. Keep in a warm place (75°F [23°C]). After 24 hours, check that fermentation has begun. Foam should be visible on the surface

See page 220 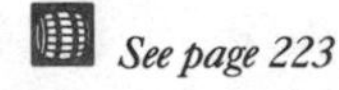 See page 223  See page 224  See page 226 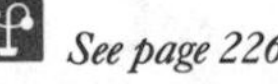 See page 227 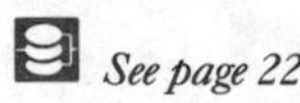 See page 229

and/or bubbles should be audible. If fermentation has not begun, see "Stuck Ferment," page 269.

8. Stir twice daily to keep the floating fruit moist.

9. Check SG every other day.

SECONDARY SEQUENCE

10. When SG reaches 1.020, scoop the cherries into a straining bag and squeeze as dry as possible into the fermentor. Discard the pulp.

11. Rack into a clean carboy. Top up with cold tap water.

12. Attach the fermentation lock.

13. Move to a cooler location, ideally 65°F (18°C).

14. After 10 days or at SG 1.000, whichever comes first, rack into a clean carboy.

15. After 3 weeks or at SG .990–.995, whichever comes first, rack into a clean carboy. Top up with cold tap water.

16. Add the finings. Top up with cold tap water. Let rest 10 days.

17. Rack into the primary fermentor.

18. Filter into a clean carboy.

19. Add ¼ teaspoon of sulphite crystals dissolved in a small amount of water. Top up with cold tap water.

20. Bulk age 1 month.

21. Add wine conditioner and bottle.

22. Bottle age 5 months.

See page 232 (col. 1)

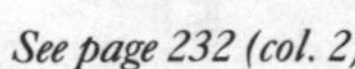
See page 232 (col. 2)

See page 238

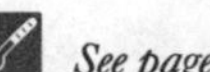
See page 240 (col. 1)

See page 240 (col. 2)

Peach Champagne

READY: 6 MONTHS

This festive wine may be consumed young and is unanimously accepted as a substitute for champagne. Should be slightly sweet.

PRIMARY INGREDIENTS

15 lb	Fresh ripe peaches	6.8 kg
1 qt	White grape concentrate	1 lit
8 lb	Sugar	3.6 kg
6 tsp	Vinacid R	
6 qt	HOT water	6 lit
2 tsp	Yeast nutrient	
2 tsp	Pectic enzyme	
1 tsp	Liquid tannin	
9	Campden tablets (crushed)	
7 qt	COLD water	7 lit
1 pkt	Champagne wine yeast	

SECONDARY INGREDIENTS

	Bentonite finings	
¼ tsp	Sulphite crystals	
9 oz	Wine conditioner	270 ml

EQUIPMENT

Basic 10 + straining bag; fine filter pads

PRIMARY SEQUENCE

1. Halve the peaches and remove the pits.

2. Place the peaches and grape concentrate in the primary fermentor. Add the hot water, sugar, and Vinacid R.

3. Stir thoroughly until all the sugar is dissolved.

4. Add the next 5 ingredients. Mix well.

5. Check, and if necessary adjust, the specific gravity (SG) of the must. It should be 1.095.

6. Check, and if necessary adjust, the temperature of the must. It should be 75°F (23°C).

7. Add the yeast to a cup of warm water. Let stand for 10 minutes. Stir in.

8. Cover the fermentor with a plastic sheet; tie down. Keep in a warm place (75°F [23°C]). After 24 hours, check that fermentation has begun. Foam should be visible on the surface

 See page 220 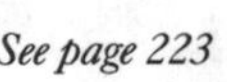  See page 223   See page 224 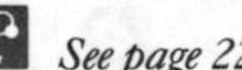  See page 226  See page 227 See page 229

and/or bubbles should be audible. If fermentation has not begun, see "Stuck Ferment," page 269.

9. Stir twice daily to keep the floating fruit moist.

10. Check SG every other day.

SECONDARY SEQUENCE

11. When SG reaches 1.020, scoop the peaches into a straining bag and squeeze the juice very gently into the fermentor so that no pulp is forced through the mesh. Discard the pulp.

12. Rack into a clean carboy. Top up with cold tap water.

13. Attach the fermentation lock.

14. Move to a cooler location, ideally 65°F (18°C).

15. After 10 days or at SG 1.000, whichever comes first, rack into a clean carboy. Top up with cold tap water.

16. After 3 weeks or at SG .990–.995, whichever comes first, rack into a clean carboy.

17. Add the finings. Top up with cold tap water. Let rest 10 days.

18. Rack into the primary fermentor.

19. Filter into a clean carboy.

20. Add ¼ teaspoon of sulphite crystals dissolved in a small amount of water. Top up with cold tap water.

21. Bulk age 1 month.

22. Sparkle the wine by the Andovin method (see "Champagne — Andovin Method," page 212) or carbonate artificially (see "Champagne — Dispatch Method," page 214).

23. Add wine conditioner and bottle.

24. Bottle age 3 months.

See page 232 (col. 1)  *See page 232 (col. 2)* 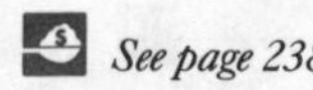 *See page 238*  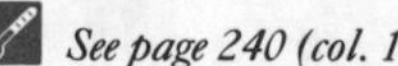 *See page 240 (col. 1)*  *See page 240 (col. 2)*

Peach Mead

READY: 1 YEAR

Peaches and honey; a wine for lovers and beekeepers

PRIMARY INGREDIENTS

15 lb	Fresh ripe peaches	6.8 kg
4 lb	Clover honey	1.8 kg
9 lb	Sugar	4 kg
6 tsp	Vinacid R	
6 qt	HOT water	6 lit
3 tsp	Yeast nutrient	
2 tsp	Pectic enzyme	
4 tsp	Liquid tannin	
8	Campden tablets (crushed)	
8 qt	COLD water	8 lit
1 pkt	Wine yeast with a high alcohol tolerance	

SECONDARY INGREDIENTS

	Bentonite finings	
¼ tsp	Sulphite crystals	
10 oz	Wine conditioner	300 ml

EQUIPMENT

Basic 10 + straining bag; fine filter pads

PRIMARY SEQUENCE

1. Halve the peaches and remove the pits.

2. Place the peaches and honey in the primary fermentor. Add the hot water, sugar, and Vinacid R.

3. Stir thoroughly until all the sugar is dissolved.

4. Add the next 5 ingredients. Mix well.

5. Check, and if necessary adjust, the specific gravity (SG) of the must. It should be 1.010.

6. Check, and if necessary adjust, the temperature of the must. It should be 75°F (23°C).

7. Add the yeast to a cup of warm water. Let stand for 10 minutes. Stir in.

8. Cover the fermentor with a plastic sheet; tie down. Keep in a warm place (75°F [23°C]). After 24 hours, check that fermentation has begun. Foam should be visible on the surface

See page 220  See page 223  See page 224 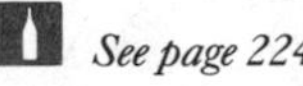  See page 226   See page 227  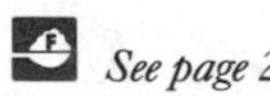 See page 229

and/or bubbles should be audible. If fermentation has not begun, see "Stuck Ferment," page 269.

9. Stir twice daily to keep the floating fruit moist.

10. Check SG every other day.

SECONDARY SEQUENCE

11. When SG reaches 1.020, scoop the peaches into a straining bag and squeeze the juice very gently into the fermentor so that no pulp is forced through the mesh. Discard the pulp.

12. Rack into a clean carboy. Top up with cold tap water.

13. Attach the fermentation lock.

14. Move to a cooler location, ideally 65°F (18°C).

15. After 10 days or at SG 1.000, whichever comes first, rack into a clean carboy. Top up with cold tap water.

16. After 3 weeks or at SG .990–.995, whichever comes first, rack into a clean carboy.

17. Add the finings. Top up with cold tap water. Let rest 10 days.

18. Rack into primary fermentor.

19. Filter into a clean carboy.

20. Add ¼ teaspoon of sulphite crystals dissolved in a small amount of water. Top up with cold tap water.

21. Bulk age 2 months.

22. Add wine conditioner and bottle.

23. Bottle age 8 months.

See page 232 (col. 1)  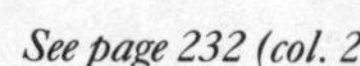 *See page 232 (col. 2)*  *See page 238*  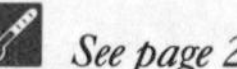 *See page 240 (col. 1)*  *See page 240 (col. 2)*

Peach Wine

READY: 7 MONTHS

A delightful fruity white wine with a low intensity of flavor; a real crowd pleaser

PRIMARY INGREDIENTS

15 lb	Fresh ripe peaches	6.8 kg
1 qt	White grape concentrate	1 lit
9 lb	Sugar	4 kg
6 qt	HOT water	6 lit
2 tsp	Yeast nutrient	
2 tsp	Pectic enzyme	
3 tsp	Liquid tannin	
8	Campden tablets (crushed)	
8 qt	COLD water	8 lit
1 pkt	Champagne wine yeast	

SECONDARY INGREDIENTS

	Claro K. C. finings	
¼ tsp	Sulphite crystals	
8 oz	Wine conditioner	240 ml

EQUIPMENT

Basic 10 + straining bag; fine filter pads

PRIMARY SEQUENCE

1. Halve the peaches and remove the pits.
2. Place the peaches and grape concentrate in the primary fermentor. Add the hot water and sugar.
3. Stir thoroughly until all the sugar is dissolved.
4. Add the next 5 ingredients. Mix well.
5. Check, and if necessary adjust, the specific gravity (SG) of the must. It should be 1.100.
6. Check, and if necessary adjust, the temperature of the must. It should be 75°F (23°C).
7. Add the yeast to a cup of warm water. Let stand for 10 minutes. Stir in.
8. Cover the fermentor with a plastic sheet; tie down. Keep in a warm place (75°F [23°C]). After 24 hours, check that fermentation has begun. Foam should be visible on the surface

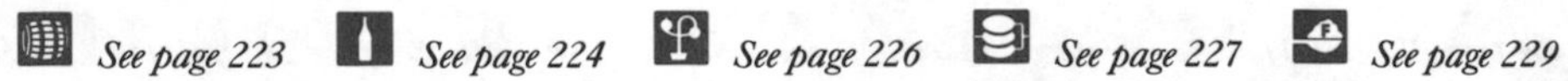
See page 220  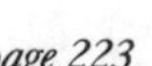 See page 223 See page 224  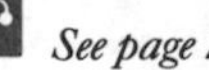 See page 226  See page 227  See page 229

and/or bubbles should be audible. If fermentation has not begun, see "Stuck Ferment," page 269.

9. Stir twice daily to keep the floating fruit moist.

10. Check SG every other day.

SECONDARY SEQUENCE

11. When SG reaches 1.020, scoop the peaches into a straining bag and squeeze the juice very gently into the fermentor so that no pulp is forced through the mesh. Discard the pulp.

12. Rack into a clean carboy. Top up with cold tap water.

13. Attach the fermentation lock.

14. Move to a cooler location, ideally 65°F (18°C).

15. After 10 days or at SG 1.000, whichever comes first, rack into a clean carboy. Top up with cold tap water.

16. After 3 weeks or at SG .990–.995, whichever comes first, rack into a clean carboy.

17. Add the finings. Top up with cold tap water. Let rest 10 days. If the wine is not clear after fining, repeat the procedure with isinglass finings. Let rest another 10 days.

18. Rack into the primary fermentor.

19. Filter into a clean carboy.

20. Add ¼ teaspoon of sulphite crystals dissolved in a small amount of water. Top up with cold tap water.

21. Bulk age 1 month.

22. Add the wine conditioner and bottle.

23. Bottle age 5 months.

 *See page 232 (col. 1)* 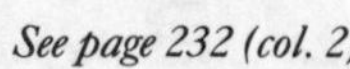 *See page 232 (col. 2)*  *See page 238*  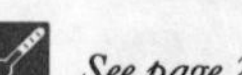 *See page 240 (col. 1)*  *See page 240 (col. 2)*

Madeira Plum Wine

READY: 1–5 YEARS

Plums make an excellent Madeira similar to sherry; a tasty aperitif that ages well.

PRIMARY INGREDIENTS

12 lb	Fresh ripe plums	5.5 kg
1 qt	White grape concentrate	1 lit
1 lb	Dried figs (chopped)	450 g
10 lb	Sugar	4.5 kg
6 qt	HOT water	6 lit
2 tsp	Yeast nutrient	
2 tsp	Pectic enzyme	
2 tsp	Liquid tannin	
8	Campden tablets (crushed)	
8 qt	COLD water	8 lit
1 pkt	Sherry yeast or wine yeast with a high alcohol tolerance	

SECONDARY INGREDIENTS

1 qt	Sugar syrup (See "Syrup Feeding," page 274.)	1 lit

(continues)

	Bentonite finings	
1 oz	Sinatin 17	30 ml
¼ tsp	Sulphite crystals	
10 oz	Wine conditioner	300 ml

EQUIPMENT

Basic 10 + fine straining bag and estufa (see "The Estufa," page 252); fine filter pads

PRIMARY SEQUENCE

1. Halve the plums and remove the pits.

2. Place the plums, chopped figs, and grape concentrate in the primary fermentor. Add the hot water and sugar.

3. Stir thoroughly until all the sugar is dissolved.

4. Add the next 5 ingredients. Mix well.

5. Check, and if necessary adjust, the temperature of the must. It should be 75°F (23°C).

6. Add the yeast to a cup of warm water. Let stand for 10 minutes. Stir in.

7. Cover the fermentor with a plastic sheet; tie down. Keep in a warm place (75°F [23°C]).

 See page 220 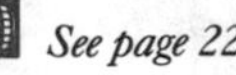 See page 223  See page 224  See page 226  See page 227  See page 229

After 24 hours, check that fermentation has begun. Foam should be visible on the surface and/or bubbles should be audible. If fermentation has not begun, see "Stuck Ferment," page 269.

8. Stir twice daily to keep the floating fruit moist.

9. Check the specific gravity (SG) every other day.

SECONDARY SEQUENCE

10. When SG reaches 1.020, scoop the plums and figs into a fine mesh straining bag and squeeze the juice very gently into the fermentor. Take care not to force any fig seeds through the mesh. Discard the pulp.

11. Rack into a clean carboy. Top up with cold tap water.

12. Attach the fermentation lock.

13. Move to a cooler location, ideally 65°F (18°C).

14. After 10 days or at SG 1.000, whichever comes first, rack into a clean carboy. Add one half of the sugar syrup. Top up with cold tap water.

15. When SG reaches 1.000, add the remaining half of the sugar syrup. Continue to ferment in a warm place until fermentation stops.

16. Add the finings. Let rest 10 days.

17. Rack into a clean carboy. Do not top up.

18. Attach the fermentation lock.

19. Place in an estufa and bake for 3 months.

20. Filter into a clean carboy.

21. Add ¼ teaspoon of sulphite crystals dissolved in a small amount of water. Add the Sinatin 17. Top up with cold tap water.

22. Bulk age 3 months.

23. Before bottling, check SG. If it is below 1.000, add the wine conditioner.

24. Bottle.

25. Bottle age a minimum of 1 year.

See page 232 (col. 1)  *See page 232 (col. 2)*  *See page 238*  *See page 240 (col. 1)*  *See page 240 (col. 2)*

Simple Plum Wine

READY: 7 MONTHS

If you have a lot of plums, they make a drinkable social wine, slightly sweet. White plums will make a better wine.

PRIMARY INGREDIENTS		
12 lb	Fresh ripe plums	5.5 kg
12 lb	Sugar	5.5 kg
4 tsp	Vinacid R	
6 qt	HOT water	6 lit
4 tsp	Liquid tannin	
2 tsp	Pectic enzyme	
2 tsp	Yeast nutrient	
8	Campden tablets (crushed)	
8 qt	COLD water	8 lit
1 pkt	Narbonne wine yeast	

SECONDARY INGREDIENTS		
	Bentonite finings	
10 oz	Wine conditioner	300 ml
¼ tsp	Sulphite crystals	

EQUIPMENT
Basic 10 + fine straining bag; fine filter pads

PRIMARY SEQUENCE

1. Halve the plums and remove the pits.

2. Place the plums in the primary fermentor. Add the hot water, sugar, and Vinacid R.

3. Stir thoroughly until all the sugar is dissolved.

4. Add the next 5 ingredients. Mix well.

5. Check, and if necessary adjust, the specific gravity (SG) of the must. It should be 1.100.

6. Check, and if necessary adjust, the temperature of the must. It should be 75°F (23°C).

7. Add the yeast to a cup of warm water. Let stand for 10 minutes. Stir in.

8. Cover the fermentor with a plastic sheet; tie down. Keep in a warm place (75°F [23°C]). After 24 hours, check that fermentation has begun. Foam should be visible on the surface and/or bubbles should be audible. If fermenta-

See page 220  See page 223  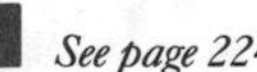 See page 224  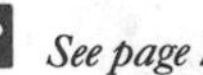 See page 226  See page 227  See page 229

tion has not begun, see “Stuck Ferment,” page 269.

9. Stir twice daily to keep the floating fruit moist.

10. Check SG every other day.

SECONDARY SEQUENCE

11. When SG reaches 1.020, scoop the plums into a straining bag and squeeze the juice very gently into the fermentor so that no pulp is forced through the mesh. Discard the pulp.

12. Rack into a clean carboy. Top up with cold tap water.

13. Attach the fermentation lock.

14. Move to a cooler location, ideally 65°F (18°C).

15. After 10 days or at SG 1.000, whichever comes first, rack into a clean carboy. Top up with cold tap water.

16. After 3 weeks or at SG .990–.995, whichever comes first, rack into a clean carboy.

17. Add the finings. Top up with cold tap water. Let rest 10 days.

18. Rack into the primary fermentor.

19. Filter into a clean carboy.

20. Add ¼ teaspoon of sulphite crystals dissolved in a small amount of water. Top up with cold tap water.

21. Bulk age 2 months.

22. Add wine conditioner and bottle.

23. Bottle age 4 months.

 *See page 232 (col. 1)* 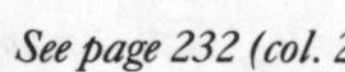 *See page 232 (col. 2)*  *See page 238*   *See page 240 (col. 1)* 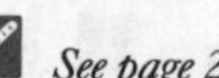  *See page 240 (col. 2)*

Summer Social Wine

READY: 7 MONTHS

A summer sipping wine made from a strawberry and cherry fruit wine base, naturally sweetened with grape concentrate

PRIMARY INGREDIENTS

96 oz	Strawberry and cherry wine base	3 lit
6 oz	Shiraz grape concentrate*	180 ml
9½ lb	Sugar	4.3 kg
10 tsp	Vinacid R	
6 qt	HOT water	6 lit
2 tsp	Yeast nutrient	
2 tsp	Pectic enzyme	
1 tsp	Liquid tannin	
5	Campden tablets (crushed)	
7 qt	COLD water	7 lit
1 pkt	Narbonne wine yeast	

SECONDARY INGREDIENTS

	Bentonite finings	
¼ tsp	Sulphite crystals	
26 oz	Shiraz grape concentrate*	780 ml
3 tsp	Potassium sorbate	

(continues)

EQUIPMENT

Basic 10 + straining bag; fine filter pads

PRIMARY SEQUENCE

1. Place the fruit base and the first 6 ounces of grape concentrate in the primary fermentor. Add the hot water, sugar, and Vinacid R.

2. Stir thoroughly until all the sugar is dissolved.

3. Add the next 5 ingredients. Mix well.

4. Check, and if necessary adjust, the specific gravity (SG) of the must. It should be 1.090.

*Purchase 1 qt (1 lit) Shiraz grape concentrate. Use 6 ounces (180 ml) in the primary ingredients and refrigerate the remainder to be used as a natural sweetener in the secondary ingredients.

 See page 220   See page 223  See page 224  See page 226 See page 227  See page 229

5. Check, and if necessary adjust, the temperature of the must. It should be 75°F (23°C).

6. Add the yeast to a cup of warm water. Let stand for 10 minutes. Stir in.

7. Cover the fermentor with a plastic sheet; tie down. Keep in a warm place (75°F [23°C]). After 24 hours, check that fermentation has begun. Foam should be visible on the surface and/or bubbles should be audible. If fermentation has not begun, see "Stuck Ferment," page 269.

8. Stir twice daily to keep the floating fruit moist.

9. Check SG every other day.

SECONDARY SEQUENCE

10. When SG reaches 1.020, scoop the fruit into a straining bag and squeeze the juice gently into the fermentor so that no pulp is forced through the mesh. Discard the pulp.

11. Rack into a clean carboy. Top up with cold tap water.

12. Attach the fermentation lock.

13. Move to a cooler location, ideally 65°F (18°C).

14. After 10 days or at SG 1.000, whichever comes first, rack into a clean carboy. Top up with cold tap water.

15. After 3 weeks or at SG .990–.995, whichever comes first, rack into a clean carboy.

16. Add the finings. Top up with cold tap water. Let rest 10 days.

17. Rack into the primary fermentor. Add the remainder of the grape concentrate.

18. Filter into a clean carboy.

19. Add ¼ teaspoon of sulphite crystals dissolved in a small amount of water. Add the potassium sorbate and stir well to mix. Top up with cold tap water.

20. Bulk age 1 month.

21. Bottle.

22. Bottle age 5 months.

See page 232 (col. 1)  *See page 232 (col. 2)*  *See page 238*  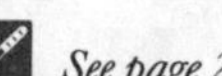 *See page 240 (col. 1)*  *See page 240 (col. 2)*

Chokecherry Wine

READY: 1 YEAR

Chokecherries are in abundance in many areas of North America, and they make a very acceptable social wine. Warning: Chokecherry stains are very difficult to remove from clothing.

PRIMARY INGREDIENTS

12 lb	Chokecherries (crushed)	5.5 kg
1 qt	Red grape concentrate	1 lit
9 lb	Sugar	4 kg
6 qt	HOT water	6 lit
2 tsp	Yeast nutrient	
2 tsp	Pectic enzyme	
8	Campden tablets (crushed)	
9 qt	COLD water	9 lit
1 pkt	Montpellier wine yeast	

SECONDARY INGREDIENTS

	Bentonite finings	
¼ tsp	Sulphite crystals	
10 oz	Wine conditioner	300 ml

EQUIPMENT

Basic 10 + straining bag and potato masher; fine filter pads

PRIMARY SEQUENCE

1. Crush the chokecherries and place them, together with the grape concentrate, in the primary fermentor. Add the hot water and sugar.

2. Stir thoroughly until all the sugar is dissolved.

3. Add the next 4 ingredients. Mix well.

4. Check, and if necessary adjust, the specific gravity (SG) of the must. It should be 1.100.

5. Check, and if necessary adjust, the temperature of the must. It should be 75°F (23°C).

6. Add the yeast to a cup of warm water. Let stand for 10 minutes. Stir in.

7. Cover the fermentor with a plastic sheet; tie down. Keep in a warm place (75°F [23°C]). After 24 hours, check that fermentation has begun. Foam should be visible on the surface

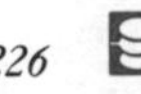
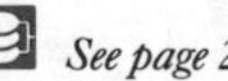

See page 220 See page 223 See page 224 See page 226 See page 227 See page 229

and/or bubbles should be audible. If fermentation has not begun, see "Stuck Ferment," page 269.

8. Stir twice daily to keep the floating fruit moist.

9. Check SG every other day.

SECONDARY SEQUENCE

10. When SG reaches 1.020, scoop the fruit into a straining bag and squeeze it dry into the fermentor. Discard the pulp.

11. Rack into a clean carboy. Top up with cold tap water.

12. Attach the fermentation lock.

13. Move to a cooler location, ideally 65°F (18°C).

14. After 10 days or at SG 1.000, whichever comes first, rack into a clean carboy. Top up with cold tap water.

15. After 3 weeks or at SG .990–.995, whichever comes first, rack into a clean carboy.

16. Add the finings. Top up with cold tap water. Let rest 10 days.

17. Rack into primary fermentor.

18. Filter into a clean carboy.

19. Add ¼ teaspoon of sulphite crystals dissolved in a small amount of water. Top up with cold tap water.

20. Bulk age 3 months.

21. Add the wine conditioner and bottle.

22. Bottle age 9 months.

See page 232 (col. 1)  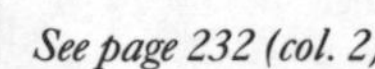 *See page 232 (col. 2)*  *See page 238*  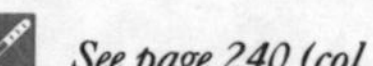 *See page 240 (col. 1)*  *See page 240 (col. 2)*

TROPICAL FRUIT WINES

QUICK REFERENCE FOR TROPICAL FRUIT WINES

Washing

Wash tropical fruit before processing.

Fruit Quality

Do not use overripe, moldy, or fermenting fruit. Remove any bruised parts. Do not use green mangoes or blackened bananas.

Preparation

Figs: chop.
Kiwi fruit: peel and chop.
Mangoes: peel and chop (discard pits).
Prickly pears: peel and chop.
Pineapple: peel and chop (discard core).
Oranges: slice but do not peel.
Bananas: chop but do not peel. Simmer in water for 30 minutes. Strain and use liquid only.

Used Pulp

Discard the pulp of tropical fruits. It cannot be used for a second run.

Banana Wine

READY: 12–15 MONTHS

Bananas make a full-bodied wine. Served over ice with a dash of soda water and a slice of lemon, it makes a great tropical cooler.

PRIMARY INGREDIENTS

12 lb	Ripe sound bananas (chopped with peel)	5.5 kg
14 lb	Sugar	6.4 kg
6 qt	HOT water	6 lit
8 tsp	Vinacid R	
4	Medium oranges (juice only)	
2 tsp	Yeast nutrient	
4 tsp	Liquid tannin	
8	Campden tablets (crushed)	
2 tsp	Pectic enzyme	
9 qt	COLD water	9 lit
1 pkt	Wine yeast with a high alcohol tolerance	

SECONDARY INGREDIENTS

1 lb	Sugar	450 g
	Bentonite finings	
10 oz	Wine conditioner	300 ml
	Vermouth flavoring (optional)	

(continues)

EQUIPMENT

Basic 10 + strainer; fine filter pads

PRIMARY SEQUENCE

1. Chop up the bananas, skin included. Place the chopped fruit in a pan with 4 quarts of water. Simmer for 30 minutes.

2. Strain off the liquid into the primary fermentor. Add the sugar, Vinacid R, and 2 quarts (2 lit) of hot tap water.

3. Stir thoroughly until all the sugar is dissolved.

4. Add the juice from 4 oranges.

5. Add the next 5 ingredients. Mix well.

6. Check, and if necessary adjust, the temperature of the must. It should be 75°F (23°C).

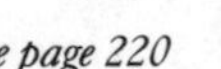
See page 220

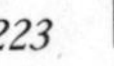
See page 223
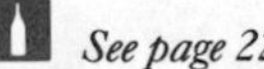
See page 224

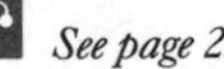
See page 226

See page 227

See page 229

7. Add the yeast to a cup of warm water. Let stand for 10 minutes. Stir in.

8. Cover the fermentor with a plastic sheet; tie down. Keep in a warm place (75°F [23°C]). After 24 hours, check that fermentation has begun. Foam should be visible on the surface and/or bubbles should be audible. If fermentation has not begun, see "Stuck Ferment," page 269.

9. Check the specific gravity (SG) every other day.

SECONDARY SEQUENCE

10. When SG reaches 1.020, rack into a clean carboy. Top up with cold tap water.

11. Attach the fermentation lock.

12. Move to a cooler location, ideally 70°F (20°C).

13. If SG drops to .990, add 1 pound (450 g) of sugar and allow to ferment for 1 month. Then rack into a clean carboy and proceed from step 14. This will give you a slightly sweet, but very strong wine. If, after 2 weeks, SG has not dropped to .990, rack into a clean carboy and proceed from step 14.

14. Add the finings. Top up with cold tap water. Let rest 10 days.

15. Rack into the primary fermentor.

16. Filter into a clean carboy.

If you did not add sugar at step 13, proceed as follows:

17. Add the wine conditioner. Add vermouth flavoring to taste. Top up with cold tap water.

18. Bulk age 3 months.

19. Bottle.

20. Bottle age 9 months.

If you added extra sugar at step 13, proceed as follows:

17. Top up with cold tap water and bulk age 3 months.

18. Add the wine conditioner. Add vermouth flavoring to taste.

19. Bottle.

20. Bottle age 1 year.

See page 232 (col. 1)  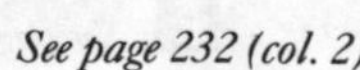 *See page 232 (col. 2)*  *See page 238*  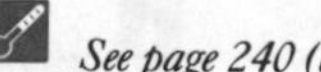 *See page 240 (col. 1)*  *See page 240 (col. 2)*

Fresh Fig Wine

READY: 1 YEAR

This wine is for those who have their very own fig tree. A pleasant social wine.

PRIMARY INGREDIENTS

24 lb	Fresh white figs (chopped)	10.9 kg
9 lb	Sugar	4 kg
15 tsp	Vinacid R	
6 qt	HOT water	6 lit
2 tsp	Yeast nutrient	
2 tsp	Pectic enzyme	
2 tsp	Liquid tannin	
8	Campden tablets (crushed)	
8 qt	COLD water	8 lit
1 pkt	Champagne wine yeast	

SECONDARY INGREDIENTS

	Bentonite finings	
¼ tsp	Sulphite crystals	
10 oz	Wine conditioner	300 ml

EQUIPMENT

Basic 10 + straining bag fine enough to contain fig seeds; fine filter pads

PRIMARY SEQUENCE

1. Chop the figs and place them in the primary fermentor. Add the hot water, sugar, and Vinacid R.

2. Stir thoroughly until all the sugar is dissolved.

3. Add the next 5 ingredients. Mix well.

4. Check, and if necessary adjust, the temperature of the must. It should be 75°F (23°C).

5. Add the yeast to a cup of warm water. Let stand for 10 minutes. Stir in.

6. Cover the fermentor with a plastic sheet; tie down. Keep in a warm place (75°F [23°C]). After 24 hours, check that fermentation has begun. Foam should be visible on the surface and/or bubbles should be audible. If fermentation has not begun, see "Stuck Ferment," page 269.

See page 220 See page 223 See page 224 See page 226 See page 227 See page 229

7. Stir twice daily to keep the floating fruit wet.

8. Check the specific gravity (SG) every other day.

SECONDARY SEQUENCE

9. When SG reaches 1.020, scoop out the figs into a very fine mesh straining bag and squeeze the juice gently into the fermentor. Take care not to force any seeds through the mesh. Discard the pulp.

10. Rack into a clean carboy. Top up with cold tap water.

11. Attach the fermentation lock.

12. Move to a cooler location, ideally 65°F (18°C).

13. After 10 days or at SG 1.000, whichever comes first, rack into a clean carboy. Top up with cold tap water.

14. After 3 weeks or at SG .990–.995, whichever comes first, rack into a clean carboy.

15. Add the finings. Top up with cold tap water. Let rest 10 days.

16. Rack into the primary fermentor.

17. Filter into a clean carboy.

18. Add ¼ teaspoon of sulphite crystals dissolved in a small amount of water. Top up with cold tap water.

19. Bulk age 2 months.

20. Add the wine conditioner and bottle.

21. Bottle age 8 months.

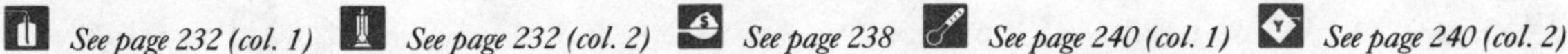
See page 232 (col. 1)

See page 232 (col. 2)

See page 238

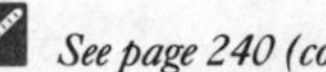
See page 240 (col. 1)

See page 240 (col. 2)

Madeira Fig Wine

READY: 2½ YEARS

Figs are an excellent ingredient for Madeira or sherry-style wines. Baking adds the characteristic sherry color and flavor.

PRIMARY INGREDIENTS

24 lb	Fresh white figs (chopped)	10.9 kg
1 qt	White grape concentrate	1 lit
9 lb	Sugar	4 kg
10 tsp	Vinacid O	
6 qt	HOT water	6 lit
2 tsp	Yeast nutrient	
2 tsp	Pectic enzyme	
2 tsp	Liquid tannin	
8	Campden tablets (crushed)	
8 qt	COLD water	8 lit
1 pkt	Sherry yeast or wine yeast with a high alcohol tolerance	

SECONDARY INGREDIENTS

	Bentonite finings	
¼ tsp	Sulphite crystals	
1 qt	Sugar syrup (See "Syrup Feeding," page 274.)	1 lit
26 oz	Rye whiskey	780 ml
10 oz	Wine conditioner	300 ml

(continues)

EQUIPMENT

Basic 10 + straining bag fine enough to contain fig seeds; estufa (see "The Estufa," page 252); fine filter pads

PRIMARY SEQUENCE

1. Chop the figs and place them, together with the grape concentrate, in the primary fermentor. Add the hot water, sugar, and Vinacid O.

2. Stir thoroughly until all the sugar is dissolved.

3. Add the next 5 ingredients. Mix well.

4. Check, and if necessary adjust, the temperature of the must. It should be 75°F (23°C).

5. Add the yeast to a cup of warm water. Let stand for 10 minutes. Stir in.

 See page 220  See page 223  See page 224  See page 226 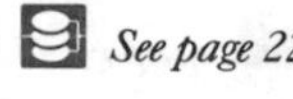 See page 227  See page 229

6. Cover the fermentor with a plastic sheet; tie down. Keep in a warm place (75°F [23°C]). After 24 hours, check that fermentation has begun. Foam should be visible on the surface and/or bubbles should be audible. If fermentation has not begun, see "Stuck Ferment," page 269.

7. Stir twice daily to keep the floating fruit wet.

8. Check the specific gravity (SG) every other day.

SECONDARY SEQUENCE

9. When SG reaches 1.020, scoop the figs into a very fine mesh straining bag and squeeze the juice gently into the fermentor. Take care not to force any seeds through the mesh. Discard the pulp.

10. Rack into a clean carboy. Top up with cold tap water.

11. Attach the fermentation lock. Continue to ferment in a warm place.

12. When SG drops to 1.000, add two cups of sugar syrup.

13. If SG drops to 1.000 again, add the remainder of the sugar syrup and rack into a clean carboy. Top up with cold tap water. If SG does not drop to 1.000, do not add more syrup, but allow the wine to ferment out.

14. When fermentation ceases, rack into a clean carboy. Top up with cold tap water.

15. After 4 weeks, add the finings. Let rest 10 days.

16. Rack into a clean carboy. Do not top up.

17. Bake in an estufa for 3 months.

18. Rack into the primary fermentor.

19. Add the rye whiskey.

20. Filter into a clean carboy.

21. Add ¼ teaspoon of sulphite crystals dissolved in a small amount of water. Sweeten to taste with wine conditioner.

22. Bottle.

23. Bottle age 2 years.

See page 232 (col. 1) 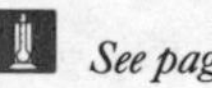  *See page 232 (col. 2)* 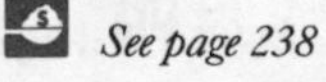 *See page 238*  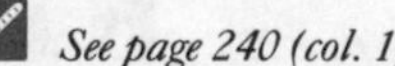 *See page 240 (col. 1)*  *See page 240 (col. 2)*

Kiwi Fruit Wine

READY: 7 MONTHS

This delicious subtropical fruit makes a light, refreshing, German-style wine.

PRIMARY INGREDIENTS

9 lb	Kiwi fruit (peeled and chopped)	4 kg
10 lb	Sugar	4.5 kg
1 qt	White grape concentrate	1 lit
15 tsp	Vinacid R	
6 qt	HOT water	6 lit
2 tsp	Yeast nutrient	
2 tsp	Pectic enzyme	
3 tsp	Liquid tannin	
8	Campden tablets (crushed)	
8 qt	COLD water	8 lit
1 pkt	Champagne wine yeast	

SECONDARY INGREDIENTS

	Claro K. C. finings	
1/4 tsp	Sulphite crystals	
8 oz	Wine conditioner	240 ml

EQUIPMENT

Basic 10 + straining bag; fine filter pads

PRIMARY SEQUENCE

1. Peel and chop the kiwi fruit and place them, together with the grape concentrate, in the primary fermentor. Add the hot water, sugar, and Vinacid R.

2. Stir thoroughly until all the sugar is dissolved.

3. Add the next 4 ingredients. Mix well.

4. Check, and if necessary adjust, the specific gravity (SG) of the must. It should be 1.095.

5. Check, and if necessary adjust, the temperature of the must. It should be 75°F (23°C).

6. Add the yeast to a cup of warm water. Let stand for 10 minutes. Stir in.

7. Cover the fermentor with a plastic sheet; tie down. Keep in a warm place (75°F [23°C]). After 24 hours, check that fermentation has begun. Foam should be visible on the surface

See page 220

See page 223

See page 224

See page 226

See page 227

See page 229

and/or bubbles should be audible. If fermentation has not begun, see "Stuck Ferment," page 269.

8. Stir twice daily to keep the floating fruit moist.

9. Check SG every other day.

SECONDARY SEQUENCE

10. When SG reaches 1.020, scoop the kiwi fruit into a straining bag and squeeze the juice gently into the fermentor. Discard the pulp.

11. Rack into a clean carboy. Top up with cold tap water.

12. Attach the fermentation lock.

13. Move to a cooler location, ideally 65°F (18°C).

14. After 10 days or at SG 1.000, whichever comes first, rack into a clean carboy. Top up with cold tap water.

15. After 3 weeks or at SG .990–.995, whichever comes first, rack into a clean carboy.

16. Add the finings. Top up with cold tap water. Let rest 10 days.

17. Rack into the primary fermentor.

18. Filter into a clean carboy.

19. Add ¼ teaspoon of sulphite crystals dissolved in a small amount of water. Top up with cold tap water.

20. Bulk age 1 month.

21. Add the wine conditioner and bottle.

22. Bottle age 5 months.

See page 232 (col. 1)

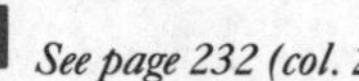
See page 232 (col. 2)
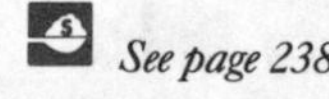
See page 238

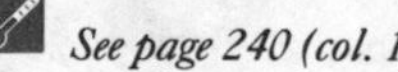
See page 240 (col. 1)

See page 240 (col. 2)

Mango Wine

READY: 7 MONTHS

This tropical fruit is recommended by U.C., Davis, for white wine. It is expensive unless mangoes grow in your area.

PRIMARY INGREDIENTS

10 lb	Mangoes (peeled and chopped)	4.5 kg
9 lb	Sugar	4 kg
1 qt	White grape concentrate	1 lit
12 tsp	Vinacid R	
6 qt	HOT water	6 lit
2 tsp	Yeast nutrient	
2 tsp	Pectic enzyme	
2 tsp	Liquid tannin	
8	Campden tablets (crushed)	
8 qt	COLD water	8 lit
1 pkt	Champagne wine yeast	

SECONDARY INGREDIENTS

	Bentonite finings	
¼ tsp	Sulphite crystals	
1 tsp	Ascorbic acid crystals	
10 oz	Wine conditioner	300 ml

(continues)

EQUIPMENT

Basic 10 + straining bag; fine filter pads

PRIMARY SEQUENCE

1. Peel and chop the mangoes and place them, together with the grape concentrate, in the primary fermentor. Add the hot water, sugar, and Vinacid R.

2. Stir thoroughly until all the sugar is dissolved.

3. Add the next 4 ingredients. Mix well.

4. Check, and if necessary adjust, the specific gravity (SG) of the must. It should be 1.095.

5. Check, and if necessary adjust, the temperature of the must. It should be 75°F (23°C).

6. Add the yeast to a cup of warm water. Let stand for 10 minutes. Stir in.

  See page 220 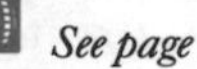 See page 223  See page 224  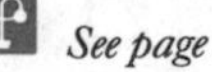 See page 226  See page 227  See page 229

7. Cover the fermentor with a plastic sheet; tie down. Keep in a warm place (75°F [23°C]). After 24 hours, check that fermentation has begun. Foam should be visible on the surface and/or bubbles should be audible. If fermentation has not begun, see "Stuck Ferment," page 269.

8. Stir twice daily to keep the floating fruit moist.

9. Check SG every other day.

SECONDARY SEQUENCE

10. When SG reaches 1.020, scoop the mangoes into a straining bag and squeeze the juice gently into the fermentor. Discard the pulp.

11. Rack into a clean carboy. Top up with cold tap water.

12. Attach the fermentation lock.

13. Move to a cooler location, ideally 65°F (18°C).

14. After 10 days or at SG 1.000, whichever comes first, rack into a clean carboy. Top up with cold tap water.

15. After 3 weeks or at SG .990–.995, whichever comes first, rack into a clean carboy.

16. Add the finings. Top up with cold tap water. Let rest 10 days.

17. Rack into the primary fermentor.

18. Filter into a clean carboy.

19. Add ¼ teaspoon of sulphite and 1 teaspoon of ascorbic acid crystals dissolved in a small amount of water. Top up with cold tap water.

20. Bulk age 1 month.

21. Add the wine conditioner and bottle.

22. Bottle age 5 months.

See page 232 (col. 1)

See page 232 (col. 2)
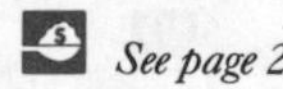
See page 238

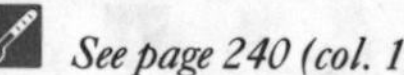
See page 240 (col. 1)

See page 240 (col. 2)

Orange and Honey Wine

READY: 1 YEAR

An exotic, slightly sweet orange mead

PRIMARY INGREDIENTS

4 lb	Mild-flavored honey	1.8 kg
4 lb	Raisins (chopped)	1.8 kg
24 oz	Frozen orange juice concentrate	710 ml
4 lb	Sugar	1.8 kg
6 tsp	Vinacid R	
6 qt	HOT water	6 lit
2 tsp	Yeast nutrient	
2 tsp	Pectic enzyme	
2 tsp	Liquid tannin	
½ oz	Dried elderflowers	15 g
8	Campden tablets (crushed)	
9 qt	COLD water	9 lit
1 pkt	Champagne wine yeast	

SECONDARY INGREDIENTS

	Claro K. C. finings	
¼ tsp	Sulphite crystals	
10 oz	Wine conditioner	300 ml

(continues)

EQUIPMENT

Basic 10 + straining bag; fine filter pads

PRIMARY SEQUENCE

1. Chop the raisins and place them, together with the honey and orange juice, in the primary fermentor. Add the hot water, sugar, and Vinacid R.

2. Stir thoroughly until all the sugar is dissolved.

3. Add the next 6 ingredients. Mix well.

4. Check, and if necessary adjust, the temperature of the must. It should be 75°F (23°C).

5. Add the yeast to a cup of warm water. Let stand for 10 minutes. Stir in.

6. Cover the fermentor with a plastic sheet; tie down. Keep in a warm place (75°F [23°C]). After 24 hours, check that fermentation has begun. Foam should be visible on the surface

See page 220  *See page 223*  *See page 224*   *See page 226*   *See page 227*  *See page 229*

and/or bubbles should be audible. If fermentation has not begun, see "Stuck Ferment," page 269.

7. Stir twice daily to keep the floating fruit moist.

8. Check the specific gravity (SG) every other day.

SECONDARY SEQUENCE

9. When SG reaches 1.020, scoop the raisins and elderflowers into a straining bag and squeeze as dry as possible into the fermentor. Discard the pulp.

10. Rack into a clean carboy. Top up with cold tap water.

11. Attach the fermentation lock.

12. Move to a cooler location, ideally 65°F (18°C).

13. After 10 days or at SG 1.000, whichever comes first, rack into a clean carboy. Top up with cold tap water.

14. After 3 weeks or at SG .990–.995, whichever comes first, rack into a clean carboy.

15. Add the finings. Top up with cold tap water. Let rest 10 days.

16. Rack into the primary fermentor.

17. Filter into a clean carboy.

18. Add ¼ teaspoon of sulphite crystals dissolved in a small amount of water. Top up with cold tap water.

19. Bulk age 3 months.

20. Add the wine conditioner and bottle.

21. Bottle age 9 months.

See page 232 (col. 1)  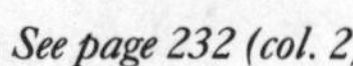 *See page 232 (col. 2)*  *See page 238*  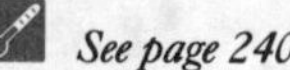 *See page 240 (col. 1)*  *See page 240 (col. 2)*

Orange Wine

READY: 1 YEAR OR MORE

Oranges ferment easily to create a high-alcohol wine. This is an aperitif wine and makes excellent vermouth.

PRIMARY INGREDIENTS

1½ qt	Frozen orange juice concentrate	1.5 lit
4	Fresh oranges (sliced with peel)	
4 lb	Sultana raisins (chopped)	1.8 kg
10 lb	Sugar	4.5 kg
6 qt	HOT water	6 lit
2 tsp	Yeast nutrient	
2 tsp	Pectic enzyme	
3 tsp	Liquid tannin	
8	Campden tablets (crushed)	
8 qt	COLD water	8 lit
1 pkt	Wine yeast with a high alcohol tolerance	

SECONDARY INGREDIENTS

	Bentonite finings	
4 cups +	Sugar syrup (See "Syrup Feeding," page 274.)	960 ml
	Vermouth flavoring (optional)	
½ tsp	Sulphite crystals	
10 oz	Wine conditioner	300 ml

EQUIPMENT

Basic 10 + straining bag; fine filter pads

PRIMARY SEQUENCE

1. Slice the oranges in their peel. Chop the raisins.

2. Place the oranges, raisins, and orange juice in the primary fermentor. Add the hot water and sugar.

3. Stir thoroughly until all the sugar is dissolved.

See page 220 See page 223 See page 224 See page 226 See page 227 See page 229

4. Add the next 5 ingredients. Mix well.

5. Check, and if necessary adjust, the specific gravity (SG) of the must. It should be 1.100–1.110.

6. Check, and if necessary adjust, the temperature of the must. It should be 75°F (23°C).

7. Add the yeast to a cup of warm water. Let stand for 10 minutes. Stir in.

8. Cover the fermentor with a plastic sheet; tie down. Keep in a warm place (75°F [23°C]). After 24 hours, check that fermentation has begun. Foam should be visible on the surface and/or bubbles should be audible. If fermentation has not begun, see "Stuck Ferment," page 269.

9. Stir twice daily to keep the floating fruit moist.

10. Check SG every other day.

SECONDARY SEQUENCE

11. When SG reaches 1.020, scoop the oranges and raisins into a straining bag and squeeze as dry as possible into the fermentor. Discard the pulp.

12. Rack into a clean carboy. Top up with cold tap water.

13. Attach the fermentation lock.

14. After 10 days or at SG 1.000, whichever comes first, rack into a clean carboy. Top up with cold tap water.

15. After 3 weeks, rack into a clean carboy. Top up with cold tap water.

16. When SG drops to .990, add 2 cups (480 ml) of sugar syrup. Attach a fermentation lock and allow to ferment in a warm place. When SG drops to .990 again, add a second 2 cups of sugar syrup. Repeat until fermentation stops.

17. Move to a cooler location, ideally 65°F (18°C).

18. Add the finings and let rest 10 days.

19. Rack into a clean carboy and add the vermouth flavoring to taste. Let rest 1 month.

20. Rack into the primary fermentor.

21. Filter into a clean carboy.

22. Add ½ teaspoon of sulphite crystals dissolved in a small amount of water. Top up with cold tap water.

23. Bulk age 3 months.

24. Sweeten to taste with wine conditioner and bottle.

25. Bottle age 8 months.

 *See page 232 (col. 1)*  *See page 232 (col. 2)*  *See page 238* 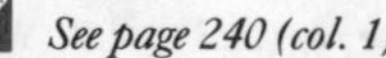 *See page 240 (col. 1)*  *See page 240 (col. 2)*

Pineapple Wine

READY: 1 YEAR

If properly aged, this wine will not have a strong pineapple flavor, but will be similar to a Sauterne.

PRIMARY INGREDIENTS

8 lb	Fresh pineapple (about 3)	3.6 kg
1 qt	White grape concentrate	1 lit
10 lb	Sugar	4.5 kg
4 tsp	Vinacid R	
6 qt	HOT water	6 lit
2 tsp	Yeast nutrient	
2 tsp	Pectic enzyme	
3 tsp	Liquid tannin	
8	Campden tablets (crushed)	
8 qt	COLD water	8 lit
1 pkt	Wine yeast with a high alcohol tolerance	

SECONDARY INGREDIENTS

2 cups	Sugar syrup (See "Syrup Feeding," page 274.)	480 ml
	Bentonite finings	
¼ tsp	Sulphite crystals	
10 oz	Wine conditioner (optional)	300 ml

(continues)

EQUIPMENT

Basic 10 + straining bag; fine filter pads

PRIMARY SEQUENCE

1. Cut the tops off the pineapples, peel them, and chop them into 1" x 1" cubes. Discard the core and overripe (brown) pieces.

2. Place the pineapple and grape concentrate in the primary fermentor. Add the hot water, sugar, and Vinacid R.

3. Stir thoroughly until all the sugar is dissolved.

4. Add the next 5 ingredients. Mix well.

5. Check, and if necessary adjust, the specific gravity (SG) of the must. It should be 1.100.

6. Check, and if necessary adjust, the temperature of the must. It should be 75°F (23°C).

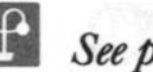

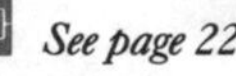

See page 220 See page 223 See page 224 See page 226 See page 227 See page 229

7. Add the yeast to a cup of warm water. Let stand for 10 minutes. Stir in.

8. Cover the fermentor with a plastic sheet; tie down. Keep in a warm place (75°F [23°C]). After 24 hours, check that fermentation has begun. Foam should be visible on the surface and/or bubbles should be audible. If fermentation has not begun, see "Stuck Ferment," page 269.

9. Stir twice daily to keep the floating fruit moist.

10. Check SG every other day.

SECONDARY SEQUENCE

11. When SG reaches 1.020, scoop the pineapple pieces into a straining bag and squeeze as dry as possible into the fermentor. Discard the pulp.

12. Rack into a clean carboy. Top up with cold tap water.

13. Attach the fermentation lock. Leave in a warm location.

14. After 10 days or at SG 1.000, whichever comes first, rack into a clean carboy. Add the sugar syrup.

15. After 3 weeks, or when fermentation ceases, move to a cooler location, ideally 65°F (18°C).

16. Add the finings and let rest 10 days.

17. Rack into the primary fermentor.

18. Filter into a clean carboy.

19. Add ¼ teaspoon of sulphite crystals dissolved in a small amount of water. Top up with cold tap water.

20. Bulk age 3 months.

21. If SG is 1.000 or less, add the wine conditioner.

22. Bottle.

23. Bottle age 9 months.

See page 232 (col. 1)  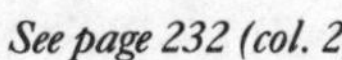 *See page 232 (col. 2)*  *See page 238*  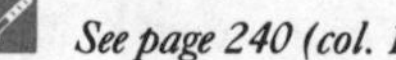 *See page 240 (col. 1)*  *See page 240 (col. 2)*

Prickly Pear Rosé

READY: 7 MONTHS

The fruit of the nopal cactus, prickly pears are a deep red in color. They make a very pleasant social wine.

PRIMARY INGREDIENTS

15 lb	Prickly pears (peeled and chopped)	6.8 kg
10 lb	Sugar	4.5 kg
15 tsp	Vinacid R	
6 qt	HOT water	6 lit
2 tsp	Yeast nutrient	
2 tsp	Pectic enzyme	
3 tsp	Liquid tannin	
8	Campden tablets (crushed)	
8 qt	COLD water	8 lit
1 pkt	Narbonne wine yeast	

SECONDARY INGREDIENTS

	Claro K. C. finings	
¼ tsp	Sulphite crystals	
10 oz	Wine conditioner	300 ml

EQUIPMENT

Basic 10 + straining bag; fine filter pads

PRIMARY SEQUENCE

1. For easier peeling, place the prickly pears in a large pan, cover them with water, and bring to a boil. Remove the pan from heat and let stand 3 minutes. Peel and chop.

2. Place the chopped prickly pears in the primary fermentor. Add the hot water, sugar, and Vinacid R.

3. Stir thoroughly until all the sugar is dissolved.

4. Add the next 5 ingredients. Mix well.

5. Check, and if necessary adjust, the specific gravity (SG) of the must. It should be 1.095.

6. Check, and if necessary adjust, the temperature of the must. It should be 75°F (23°C).

7. Add the yeast to a cup of warm water. Let stand for 10 minutes. Stir in.

8. Cover the fermentor with a plastic sheet; tie down. Keep in a warm place (75°F [23°C]).

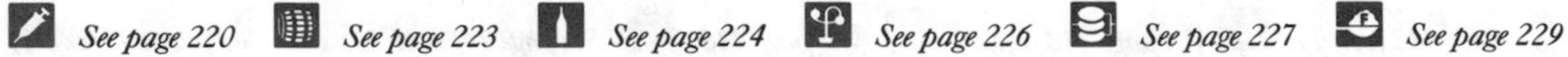
See page 220  See page 223  See page 224  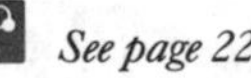 See page 226  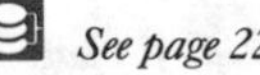 See page 227  See page 229

After 24 hours, check that fermentation has begun. Foam should be visible on the surface and/or bubbles should be audible. If fermentation has not begun, see "Stuck Ferment," page 269.

9. Stir twice daily to keep the floating fruit moist.

10. Check SG every other day.

SECONDARY SEQUENCE

11. When SG reaches 1.020, scoop the prickly pears into a straining bag and squeeze the juice gently into the fermentor. Discard the pulp.

12. Rack into a clean carboy. Top up with cold tap water.

13. Attach the fermentation lock.

14. Move to a cooler location, ideally 65°F (18°C).

15. After 10 days or at SG 1.000, whichever comes first, rack into a clean carboy. Top up with cold tap water.

16. After 3 weeks or at SG .990–.995, whichever comes first, rack into a clean carboy.

17. Add the finings. Top up with cold tap water. Let rest 10 days.

18. Rack into the primary fermentor.

19. Filter into a clean carboy.

20. Add ¼ teaspoon of sulphite crystals dissolved in a small amount of water. Top up with cold tap water.

21. Bulk age 1 month.

22. Add the wine conditioner and bottle.

23. Bottle age 5 months.

 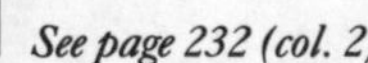 *See page 232 (col. 1)*  *See page 232 (col. 2)*  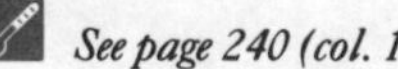 *See page 238*  *See page 240 (col. 1)* *See page 240 (col. 2)*

DRIED FRUIT WINES

QUICK REFERENCE FOR DRIED FRUIT WINES

Fruit Quality

Use freshly packaged dried fruits. Avoid fruit with any sign of mold.

Preparation

Chop raisins, figs, dates (discard pits), and dried wine grapes.

Specific Gravity

It is difficult to take an accurate specific gravity reading, because dates, raisins, and figs are 60% sugar, which is slowly released during fermentation.

Used Pulp

Discard the pulp of dried fruits. It cannot be used for a second run.

Date Wine

READY: 1 YEAR

A pleasant social wine that will keep for years

PRIMARY INGREDIENTS		
10 lb	Dried dates (chopped)	4.5 kg
6 lb	Sugar	2.7 kg
3	Oranges (sliced)	
15 tsp	Vinacid R	
6 qt	HOT water	6 lit
2 tsp	Yeast nutrient	
2 tsp	Liquid tannin	
1 tsp	Pectic enzyme	
8	Campden tablets (crushed)	
9 qt	COLD water	9 lit
1 pkt	Champagne wine yeast	

SECONDARY INGREDIENTS		
2 cups	Sugar syrup (See "Syrup Feeding," page 274.)	480 ml
	Claro K. C. finings	
¼ tsp	Sulphite crystals	

EQUIPMENT

Basic 10 + straining bag; fine filter pads

PRIMARY SEQUENCE

1. Chop the dates (remove the pits) and place them in the primary fermentor. Add the sliced oranges, hot water, sugar, and Vinacid R.

2. Stir thoroughly until all the sugar is dissolved.

3. Add the next 5 ingredients. Mix well.

4. Check, and if necessary adjust, the temperature of the must. It should be 75°F (23°C).

5. Add the yeast to a cup of warm water. Let stand for 10 minutes. Stir in.

6. Cover the fermentor with a plastic sheet; tie down. Keep in a warm place (75°F [23°C]). After 24 hours, check that fermentation has begun. Foam should be visible on the surface and/or bubbles should be audible. If fermentation has not begun, see "Stuck Ferment," page 269.

See page 220  *See page 223* *See page 224*  *See page 226*  *See page 227*  *See page 229*

7. Stir twice daily to keep the floating fruit moist.

8. Check the specific gravity (SG) every other day.

SECONDARY SEQUENCE

9. When SG reaches 1.020, scoop the dates and oranges into a straining bag and squeeze the juice gently into the fermentor. Discard the pulp.

10. Rack into a clean carboy. Do not top up with cold tap water.

11. Attach fermentation lock. Continue to ferment in a warm place.

12. When SG falls below 1.000, top up with sugar syrup.

13. If SG falls below 1.000 again, add the remainder of the syrup and continue to ferment in a warm place. If SG does not fall below 1.000 again, do not add the remainder of the syrup.

14. When the fermentation stops, rack into a clean carboy.

15. Add the finings and let rest 10 days.

16. Rack into the primary fermentor.

17. Filter into a clean carboy.

18. Add ¼ teaspoon of sulphite crystals dissolved in a small amount of water.

19. Bulk age 3 months.

20. Bottle.

21. Bottle age 9 months.

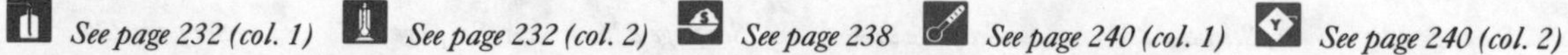

See page 232 (col. 1)

See page 232 (col. 2)

See page 238
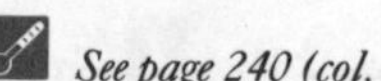
See page 240 (col. 1)

See page 240 (col. 2)

Dried Fig Wine

READY: 11 MONTHS

The simplest way to make a sherry-style wine without baking in an estufa.

PRIMARY INGREDIENTS

4 lb	Dried figs (chopped)	1.8 kg
2	Oranges (sliced)	
10 lb	Sugar	4.5 kg
15 tsp	Vinacid R	
6 qt	HOT water	6 lit
2 tsp	Yeast nutrient	
2 tsp	Liquid tannin	
8	Campden tablets (crushed)	
9 qt	COLD water	9 lit
1 pkt	Sherry wine yeast	

SECONDARY INGREDIENTS

	Claro K. C. finings	
1 tbsp	Sinatin 17	
12 oz	Wine conditioner	354 ml
¼ tsp	Sulphite crystals	

EQUIPMENT

Basic 10 + straining bag; fine filter pads

PRIMARY SEQUENCE

1. Chop the figs and place them in the primary fermentor. Add the sliced oranges, sugar, hot water, and Vinacid R.

2. Stir thoroughly until all the sugar is dissolved.

3. Add the next 4 ingredients and mix well.

4. Check, and if necessary adjust, the temperature of the must. It should be 75°F (23°C).

5. Add the yeast to a cup of warm water. Let stand 10 minutes. Stir in.

6. Cover the fermentor with a plastic sheet; tie down. Keep in a warm place (75°F [23°C]). After 24 hours, check that fermentation has begun. Foam should be visible on the surface and/or bubbles should be audible. If fermentation has not begun, see "Stuck Ferment," page 269.

7. Stir twice daily to keep the floating fruit moist.

 See page 220  See page 223 See page 224  See page 226  See page 227  See page 229

8. Check the specific gravity (SG) every other day.

SECONDARY SEQUENCE

9. When SG reaches 1.020, scoop the figs and oranges into a straining bag and squeeze the juice gently into the fermentor. Discard the pulp.

10. Rack into a clean carboy. Top up with cold tap water.

11. Attach the fermentation lock.

12. Move to cooler location, ideally 68°F (20°C).

13. After 10 days, or at SG 1.000, whichever comes first, rack into a clean carboy. Top up with cold tap water.

14. After 3 weeks or at SG .990–.995, whichever comes first, rack into a clean carboy.

15. Add finings. Top up with cold tap water. Let rest 10 days.

16. Rack into the primary fermentor.

17. Filter into a clean carboy.

18. Add ¼ teaspoon of sulphite crystals dissolved in a small amount of water. Add the Sinatin 17. Top up with cold tap water.

19. Bulk age 1 month.

20. Bottle.

21. Bottle age 9 months.

See page 232 (col. 1)  *See page 232 (col. 2)* 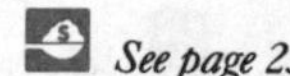 *See page 238* 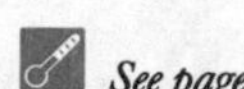 *See page 240 (col. 1)* 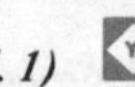 *See page 240 (col. 2)*

Raisin Wine

READY: 7 MONTHS

Raisins are dried by dehydrators or in the sun. They are usually sultanas or Thompson seedless. They make an excellent social wine with a golden-to-amber color.

PRIMARY INGREDIENTS

12 lb	Raisins (chopped)	5.5 kg
8 lb	Sugar	3.6 kg
16 tsp	Vinacid R	
6 qt	HOT water	6 lit
2 tsp	Yeast nutrient	
4 tsp	Liquid tannin	
2 tsp	Pectic enzyme	
8	Campden tablets (crushed)	
9 qt	COLD water	9 lit
1 pkt	Wine yeast with a high alcohol tolerance	

SECONDARY INGREDIENTS

	Bentonite finings	
8 oz	Wine conditioner	240 ml
¼ tsp	Sulphite crystals	

EQUIPMENT

Basic 10 + straining bag; fine filter pads

PRIMARY SEQUENCE

1. Chop the raisins and place them in the primary fermentor. Add the hot water, sugar, and Vinacid R.

2. Stir thoroughly until all the sugar is dissolved.

3. Add the next 5 ingredients. Mix well.

4. Check, and if necessary adjust, the temperature of the must. It should be 75°F (23°C).

5. Add the yeast to a cup of warm water. Let stand for 10 minutes. Stir in.

6. Cover the fermentor with a plastic sheet; tie down. Keep in a warm place (75°F [23°C]). After 24 hours, check that fermentation has begun. Foam should be visible on the surface and/or bubbles should be audible. If fermentation has not begun, see "Stuck Ferment," page 269.

7. Stir twice daily to keep the floating fruit moist.

 See page 220  See page 223  See page 224   See page 226  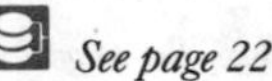 See page 227  See page 229

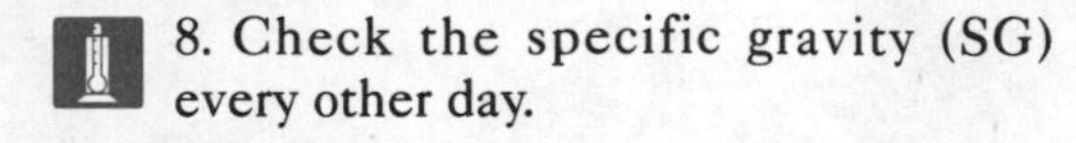

8. Check the specific gravity (SG) every other day.

SECONDARY SEQUENCE

9. When SG reaches 1.020, scoop the raisins into a straining bag and squeeze it as dry as possible into the fermentor. Discard the pulp.

10. Rack into a clean carboy. Top up with cold tap water.

11. Attach the fermentation lock.

12. Move to a cooler location, ideally 65°F (18°C).

13. After 10 days or at SG 1.000, whichever comes first, rack into a clean carboy. Top up with cold tap water.

14. After 3 weeks or at SG .990–.995, whichever comes first, rack into a clean carboy.

15. Add the finings. Top up with cold tap water. Let rest 10 days.

16. Rack into the primary fermentor.

17. Filter into a clean carboy.

18. Add ¼ teaspoon of sulphite crystals dissolved in a small amount of water. Top up with cold tap water.

19. Bulk age 3 months.

20. Add wine conditioner and bottle.

21. Bottle age 3 months.

See page 232 (col. 1)  *See page 232 (col. 2)*  *See page 238* *See page 240 (col. 1)*  *See page 240 (col. 2)*

Raisin Dessert Wine

READY: 1 YEAR

This is a strong, sweet sipping wine for after dinner or before bedtime.

PRIMARY INGREDIENTS

12 lb	Raisins (chopped)	5.5 kg
8 lb	Sugar	3.6 kg
16 tsp	Vinacid O	
6 qt	HOT water	6 lit
2 tsp	Yeast nutrient	
2 tsp	Pectic enzyme	
4 tsp	Liquid tannin	
8	Campden tablets (crushed)	
9 qt	COLD water	9 lit
1 pkt	Wine yeast with a high alcohol tolerance	

SECONDARY INGREDIENTS

1 qt	Sugar syrup (See "Syrup Feeding," page 274.)	1 lit
	Bentonite finings	
¼ tsp	Sulphite crystals	
8 oz	Wine conditioner (optional)	240 ml

(continues)

EQUIPMENT

Basic 10 + straining bag; fine filter pads

PRIMARY SEQUENCE

1. Chop the raisins and place them in the primary fermentor. Add the hot water, sugar, and Vinacid O.

2. Stir thoroughly until all the sugar is dissolved.

3. Add the next 5 ingredients. Mix well.

4. Check, and if necessary adjust, the temperature of the must. It should be 75°F (23°C).

5. Add the yeast to a cup of warm water. Let stand for 10 minutes. Stir in.

6. Cover the fermentor with a plastic sheet; tie down. Keep in a warm place (75°F [23°C]). After 24 hours, check that fermentation has begun. Foam should be visible on the surface

See page 220  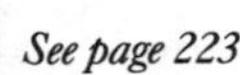 See page 223 See page 224  See page 226  See page 227  See page 229

and/or bubbles should be audible. If fermentation has not begun, see “Stuck Ferment,” page 269.

7. Stir twice daily to keep the floating fruit moist.

8. Check the specific gravity (SG) every other day.

SECONDARY SEQUENCE

9. When SG reaches 1.020, scoop the raisins into a straining bag and squeeze it as dry as possible into the fermentor. Discard the pulp.

10. Rack into a clean carboy. Do not top up.

11. Attach the fermentation lock. Continue to ferment in a warm place.

12. When SG falls below 1.000, top up with sugar syrup. Continue to ferment in a warm place.

13. If SG falls below 1.000 again, add the remainder of the syrup. If SG does not fall below 1.000 again, do not add the remainder of the syrup.

14. When the fermentation stops, rack into a clean carboy. Top up with cold tap water.

15. Add the finings and let rest for 10 days.

16. Rack into the primary fermentor.

17. Filter into a clean carboy.

18. Add ¼ teaspoon of sulphite crystals dissolved in a small amount of water. Top up with cold tap water.

19. Bulk age 3 months.

20. Sweeten to taste with wine conditioner and bottle.

21. Bottle age 9 months. This wine will keep for years.

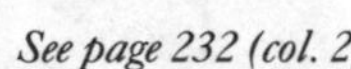

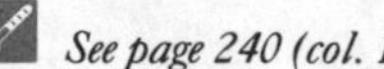

See page 232 (col. 1) *See page 232 (col. 2)* *See page 238* *See page 240 (col. 1)* *See page 240 (col. 2)*

Rice Wine

READY: 7 MONTHS

This is not a true sake, but a raisin wine. It makes a slightly sweet social wine.

PRIMARY INGREDIENTS

10 lb	Raisins (chopped)	4.5 kg
5 lb	Rice (regular white)	2.2 kg
10 lb	Sugar	4.5 kg
15 tsp	Vinacid R	
6 qt	HOT water	6 lit
4 tsp	Liquid tannin	
2 tsp	Pectic enzyme	
2 tsp	Yeast nutrient	
8	Campden tablets (crushed)	
8 qt	COLD water	8 lit
1 pkt	Champagne wine yeast	

SECONDARY INGREDIENTS

	Bentonite finings	
¼ tsp	Sulphite crystals	
8 oz	Wine conditioner	240 ml

EQUIPMENT

Basic 10 + straining bag; fine filter pads

PRIMARY SEQUENCE

1. Chop the raisins and place them, together with the rice, in the primary fermentor. Add the hot water, sugar, and Vinacid R.

2. Stir thoroughly until all the sugar is dissolved.

3. Add the next 5 ingredients. Mix well.

4. Check, and if necessary adjust, the temperature of the must. It should be 75°F (23°C).

5. Add the yeast to a cup of warm water. Let stand for 10 minutes. Stir in.

6. Cover the fermentor with a plastic sheet; tie down. Keep in a warm place (75°F [23°C]). After 24 hours, check that fermentation has begun. Foam should be visible on the surface and/or bubbles should be audible. If fermentation has not begun, see "Stuck Ferment," page 269.

7. Stir twice daily to keep the floating fruit moist.

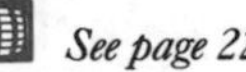

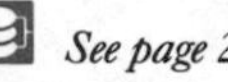

See page 220 See page 223 See page 224 See page 226 See page 227 See page 229

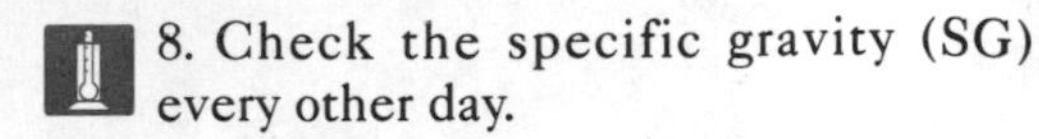

8. Check the specific gravity (SG) every other day.

SECONDARY SEQUENCE

9. When SG reaches 1.020, scoop the raisins and rice into a straining bag and squeeze it as dry as possible into the fermentor. Discard the pulp.

10. Rack into a clean carboy. Top up with cold tap water.

11. Attach the fermentation lock.

12. Move to a cooler location, ideally 65°F (18°C).

13. After 10 days or at SG 1.000, whichever comes first, rack into a clean carboy. Top up with cold tap water.

14. After 3 weeks, rack into a clean carboy. Top up with cold tap water.

15. When fermentation stops, add the finings. Let rest 10 days.

16. Rack into the primary fermentor.

17. Filter into a clean carboy.

18. Add ¼ teaspoon of sulphite crystals dissolved in a small amount of water. Top up with cold tap water.

19. Bulk age 1 month.

20. Add the wine conditioner only if SG is below 1.000.

21. Bottle.

22. Bottle age 5 months. This wine will keep for years.

See page 232 (col. 1) *See page 232 (col. 2)* *See page 238* *See page 240 (col. 1)* *See page 240 (col. 2)*

Table Wine from Dried Vinifera Grapes

READY: 16 MONTHS

Dried Shiraz wine grapes from Australia are very convenient; they have a long shelf life and ship well when purchased by mail order.

PRIMARY INGREDIENTS

12 lb	Dried Shiraz wine grapes (chopped)	5.5 kg
6 lb	Sugar	2.7 kg
14 tsp	Vinacid O	
6 qt	HOT water	6 lit
2 tsp	Yeast nutrient	
2 tsp	Pectic enzyme	
9 qt	COLD water	9 lit
1 pkt	Montpellier wine yeast	

SECONDARY INGREDIENTS

	Bentonite finings	
¼ tsp	Sulphite crystals	

EQUIPMENT

Basic 10 + large straining bag; coarse filter pads

PRIMARY SEQUENCE

1. Chop the dried grapes and place them in the primary fermentor. Add the hot water, sugar, and Vinacid O.

2. Stir thoroughly until all the sugar is dissolved.

3. Add the next 3 ingredients. Mix well.

4. Check, and if necessary adjust, the specific gravity (SG) of the must. It should be 1.080.

5. Check, and if necessary adjust, the temperature of the must. It should be 75°F (23°C).

6. Add the yeast to a cup of warm water. Let stand for 10 minutes. Stir in.

7. Cover the fermentor with a plastic sheet; tie down. Keep in a warm place (75°F [23°C]). After 24 hours, check that fermentation has begun. Foam should be visible on the surface and/or bubbles should be audible. If fermentation has not begun, see "Stuck Ferment," page 269.

8. Stir twice daily to keep the floating fruit moist.

9. Check the SG every other day.

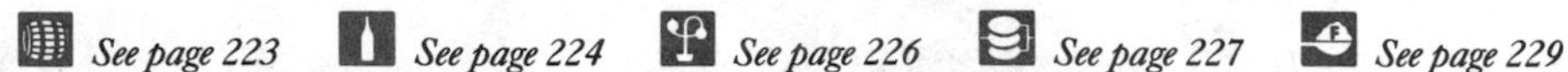
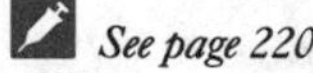

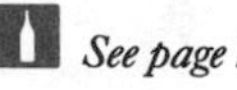

See page 220 See page 223 See page 224 See page 226 See page 227 See page 229

SECONDARY SEQUENCE

10. When SG reaches 1.010, scoop the grapes into a straining bag and squeeze it as dry as possible into the fermentor. Discard the pulp.

11. Rack into a clean carboy. Top up with cold tap water.

12. Attach the fermentation lock.

13. Move to a cooler location, ideally 65°F (18°C).

14. After 10 days or at SG 1.000, whichever comes first, rack into a clean carboy. Top up with cold tap water.

15. After 3 weeks or at SG .990–.995, whichever comes first, rack into a clean carboy.

16. Add the finings. Top up with cold tap water. Let rest 10 days.

17. Rack into a clean carboy. Top up with cold tap water. Let rest 3 months.

18. Filter into a clean carboy.

19. Add ¼ teaspoon of sulphite crystals dissolved in a small amount of water. Top up with cold tap water.

20. Bulk age 3 months.

21. Bottle.

22. Bottle age 9 months.

See page 232 (col. 1)  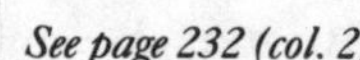 *See page 232 (col. 2)*  *See page 238*  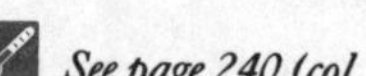 *See page 240 (col. 1)*  *See page 240 (col. 2)*

GRAPE CONCENTRATE WINES

QUICK REFERENCE FOR GRAPE CONCENTRATE RECIPES

Quality

There are two products on the market currently labeled grape concentrate. One is pure grape concentrate with no additives except the minimum amount of sulphur dioxide to prevent spoilage. The other is a mixture of grape and/or pear concentrate, citric acid, and sugar, and is a wine base rather than a grape concentrate. The latter is cheaper because it seldom contains the higher-priced varietal grapes; nevertheless it is often labeled Burgundy, Bordeaux, Chablis, and so on. Both products entail the same amount of work; but whereas the pure grape concentrate will make excellent wine, the wine base will make consistently poor, thin wine. Always read the list of ingredients on the container when buying concentrate.

Acid Content

With pure grape concentrate, the acid content will vary with each harvest. Ask your supplier about the acid content of the concentrate you are purchasing and add the amount of acid he recommends, or perform the acid test we describe (see "Acidity," page 220) and make your own adjustment. Remember that the acid content of grape concentrate musts increases slightly during fermentation. Therefore, begin red wines at 4 g/l and white wines at 5 g/l.

Tannin

Grape concentrate is grape juice with 60–70% of the water removed. To use concentrate, many recipes recommend restoring this water and leave it at that. While this may sound logical, the results are often disappointing. Concentrate contains very little tannin because it has not been fermented, and it is during the fermentation process that tannin is extracted from skins, seeds, and stems. For this reason, our recipes specify significantly more tannin than most conventional recipes.

Remember, liquid tannin differs in strength from tannin in powder form. If the recipe calls for 1 teaspoon of liquid tannin, and liquid tannin is not available, use 2 teaspoons of dried tannin.

Barolo

READY: 6 MONTHS +

A full-bodied, rich, Italian wine. Consumed young, it is a tannic, rough wine for rich, spicy food. Well aged, it becomes full, soft, almost sweet. To savor the potential of this wine, set some bottles aside and allow them to mature.

PRIMARY INGREDIENTS		
4 qt	Italian grape concentrate from the Parma region	4 lit
6 qt	HOT water	6 lit
3 lb	Sugar	1.3 kg
2½ tsp	Vinacid O	
1 lb	Dried Shiraz wine grapes (coarsely chopped)	450 g
2 tsp	Liquid tannin	
1 tsp	Yeast nutrient	
7 qt	COLD water	7 lit
1 pkt	Wine yeast (Montpellier preferred)	

SECONDARY INGREDIENTS		
1½ oz	Sinatin 17 or oak sticks	45 ml
	Claro K. C. finings	

EQUIPMENT
Basic 10; coarse filter pads

PRIMARY SEQUENCE

1. Place the grape concentrate in the primary fermentor. Add the hot water, sugar, and Vinacid O.

2. Stir thoroughly until all the sugar is dissolved.

3. Chop the grapes coarsely and add to the fermentor. Mix well.

4. Add the next 3 ingredients. Mix well.

5. Check, and if necessary adjust, the temperature of the must. It should be 75°F (23°C).

6. Add the yeast to a cup of warm water. Let stand for 10 minutes. Stir in.

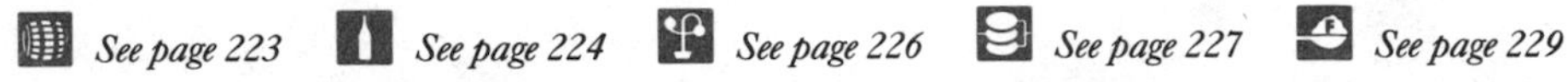
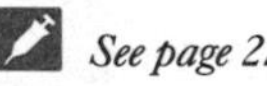
See page 220

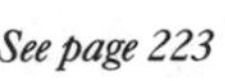
See page 223

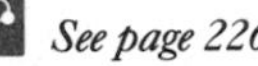
See page 224

See page 226
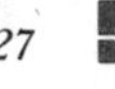
See page 227
See page 229

7. Cover the fermentor with a plastic sheet; tie down. Keep in a warm place (75°F [23°C]). After 24 hours, check that fermentation has begun. Foam should be visible on the surface and/or bubbles should be audible. If fermentation has not begun, see "Stuck Ferment," page 269.

8. Stir twice daily to keep the floating fruit moist.

9. Check the specific gravity (SG) every other day.

SECONDARY SEQUENCE

10. When SG reaches 1.020, rack into a clean carboy, leaving the grapes behind in the fermentor. Discard the grapes. Top up with cold tap water.

11. Attach the fermentation lock.

12. Move to a cooler location, ideally 65°F (18°C).

13. After 10 days or at SG 1.000, whichever comes first, rack into a clean carboy. Top up with cold tap water.

14. After 3 weeks or at SG .990–.995, whichever comes first, rack into a clean carboy.

15. Add the finings. Top up with cold tap water. Let rest 10 days.

16. Rack into the primary fermentor.

17. Filter into a clean carboy. Top up with cold tap water.

18. Bulk age 6 months.

19. Bottle.

20. Bottle age 6 months.

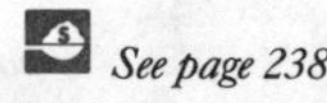

See page 232 (col. 1) *See page 232 (col. 2)* *See page 238* *See page 240 (col. 1)* *See page 240 (col. 2)*

Bordeaux-Style Red Wine

READY: 16 MONTHS

Australia's best-known blend of wine grapes, Malbec and Shiraz. Deserves some cellar time.

PRIMARY INGREDIENTS

3 qt	Malbec grape concentrate	3 lit
1 qt	Shiraz grape concentrate	1 lit
3 lb	Sugar	1.3 kg
4 tsp	Vinacid O	
6 qt	HOT water	6 lit
1 tsp	Yeast nutrient	
2 tsp	Pectic enzyme	
4 tsp	Liquid tannin	
8 qt	COLD water	8 lit
1 pkt	Montpellier wine yeast	

SECONDARY INGREDIENTS

	Claro K. C. finings	
¼ tsp	Sulphite crystals	
1½ oz	Sinatin 17	45 ml

EQUIPMENT

Basic 10; coarse filter pads

PRIMARY SEQUENCE

1. Place the grape concentrates in the primary fermentor. Add the hot water, sugar, and Vinacid O.

2. Stir thoroughly until all the sugar is dissolved.

3. Add the next 4 ingredients. Mix well.

4. Check, and if necessary adjust, the specific gravity (SG) of the must. It should be 1.100.

5. Check, and if necessary adjust, the temperature of the must. It should be 75°F (23°C).

6. Add the yeast to a cup of warm water. Let stand for 10 minutes. Stir in.

7. Cover the fermentor with a plastic sheet; tie down. Keep in a warm place (75°F [23°C]). After 24 hours, check that fermentation has begun. Foam should be visible on the surface and/or bubbles should be audible. If

See page 220   See page 223 See page 224   See page 226  See page 227  See page 229

fermentation has not begun, see "Stuck Ferment," page 269.

8. Check the SG every other day.

SECONDARY SEQUENCE

9. When SG reaches 1.020, rack into a clean carboy. Top up with cold tap water.

10. Attach the fermentation lock.

11. Move to a cooler location, ideally 65°F (18°C).

12. After 10 days or at SG 1.000, whichever comes first, rack into a clean carboy. Top up with cold tap water.

13. After 3 weeks or at SG .990–.995, whichever comes first, rack into a clean carboy.

14. Add the finings. Top up with cold tap water. Let rest 10 days.

15. Rack into the primary fermentor.

16. Filter into a clean carboy.

17. Add ¼ teaspoon of sulphite crystals dissolved in a small amount of water. Add the Sinatin 17. Top up with cold tap water.

18. Bulk age 3 months.

19. Bottle.

20. Bottle age 12 months.

See page 232 (col. 1)  *See page 232 (col. 2)*  *See page 238*  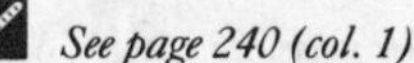 *See page 240 (col. 1)*  *See page 240 (col. 2)*

Cabernet Sauvignon

READY: 1 YEAR

The king of grapes, used in the great Bordeaux wines

PRIMARY INGREDIENTS

3 qt	Cabernet Sauvignon grape concentrate	3 lit
1 lb	Dried wine grapes (chopped)	450 g
3½ lb	Sugar	1.6 kg
4 tsp	Vinacid O	
6 qt	HOT water	6 lit
1 tsp	Yeast nutrient	
2 tsp	Pectic enzyme	
2 tsp	Liquid tannin	
9 qt	COLD water	9 lit
1 pkt	Montpellier wine yeast	

SECONDARY INGREDIENTS

	Claro K. C. finings	
1 oz	Sinatin 17	30 ml
¼ tsp	Sulphite crystals	

EQUIPMENT

Basic 10 + straining bag; coarse filter pads

PRIMARY SEQUENCE

1. Chop the dried grapes and place them, together with the grape concentrate, in the primary fermentor. Add the hot water, sugar, and Vinacid O.

2. Stir thoroughly until all the sugar is dissolved.

3. Add the next 4 ingredients. Mix well.

4. Check, and if necessary adjust, the specific gravity (SG) of the must. It should be 1.100.

5. Check, and if necessary adjust, the temperature of the must. It should be 75°F (23°C).

6. Add the yeast to a cup of warm water. Let stand for 10 minutes. Stir in.

7. Cover the fermentor with a plastic sheet; tie down. Keep in a warm place (75°F [23°C]). After 24 hours, check that fermentation has begun. Foam should be visible on the surface

See page 220

See page 223

See page 224

See page 226

See page 227

See page 229

and/or bubbles should be audible. If fermentation has not begun, see "Stuck Ferment," page 269.

8. Stir twice daily to keep the floating fruit moist.

9. Check the SG every other day.

SECONDARY SEQUENCE

10. When SG reaches 1.020, scoop the grapes into a straining bag and squeeze as dry as possible into the fermentor. Discard the pulp.

11. Rack into a clean carboy. Top up with cold tap water.

12. Attach the fermentation lock.

13. Move to a cooler location, ideally 65°F (18°C).

14. After 10 days or at SG 1.000, whichever comes first, rack into a clean carboy. Top up with cold tap water.

15. After 3 weeks or at SG .990–.995, whichever comes first, rack into a clean carboy.

16. Add the finings. Top up with cold tap water. Let rest 10 days.

17. Rack into the primary fermentor. Add Sinatin 17.

18. Filter into a clean carboy.

19. Add ¼ teaspoon of sulphite crystals dissolved in a small amount of water. Top up with cold tap water.

20. Bulk age 3 months.

21. Bottle.

22. Bottle age 9 months.

See page 232 (col. 1)  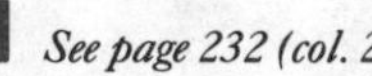 *See page 232 (col. 2)*  *See page 238*   *See page 240 (col. 1)*  *See page 240 (col. 2)*

Chianti

READY: 1 YEAR

A well-balanced, full-bodied wine; goes well with Italy's rich, flavorful pastas.

EQUIPMENT

Basic 10 + straining bag; coarse filter pads

PRIMARY INGREDIENTS

4½ qt	Italian Chianti grape concentrate	4.3 lit
1 lb	Dried wine grapes (chopped)	450 g
3¼ lb	Sugar	1.5 kg
4 tsp	Vinacid O	
6 qt	HOT water	6 lit
1 tsp	Yeast nutrient	
1 tsp	Liquid tannin	
8 qt	COLD water	8 lit
1 pkt	Montpellier wine yeast	

SECONDARY INGREDIENTS

	Claro K. C. finings	
¼ tsp	Sulphite crystals	

PRIMARY SEQUENCE

1. Place the grape concentrate and the dried grapes in the primary fermentor. Add the hot water, sugar, and Vinacid O.

2. Stir thoroughly until all the sugar is dissolved.

3. Add the next 3 ingredients. Mix well.

4. Check, and if necessary adjust, the specific gravity (SG) of the must. It should be 1.100.

5. Check, and if necessary adjust, the temperature of the must. It should be 75°F (23°C).

6. Add the yeast to a cup of warm water. Let stand for 10 minutes. Stir in.

7. Cover the fermentor with a plastic sheet; tie down. Keep in a warm place (75°F [23°C]). After 24 hours, check that fermentation has begun. Foam should be visible on the surface and/or bubbles should be audible. If

See page 220  See page 223  See page 224  See page 226  See page 227  See page 229

fermentation has not begun, see "Stuck Ferment," page 269.

8. Stir twice daily to keep the floating fruit moist.

9. Check SG every other day.

SECONDARY SEQUENCE

10. When SG reaches 1.020, scoop the grapes into a straining bag and squeeze it as dry as possible into the fermentor. Discard the pulp.

11. Rack into a clean carboy. Top up with cold tap water.

12. Attach the fermentation lock.

13. Move to a cooler location, ideally 65°F (18°C).

14. After 10 days or at SG 1.000, whichever comes first, rack into a clean carboy. Top up with cold tap water.

15. After 3 weeks or at SG .990–.995, whichever comes first, rack into a clean carboy.

16. Add the finings. Top up with cold tap water. Let rest 10 days.

17. Rack into the primary fermentor.

18. Filter into a clean carboy.

19. Add ¼ teaspoon of sulphite crystals dissolved in a small amount of water. Top up with cold tap water.

20. Bulk age 3 months.

21. Bottle.

22. Bottle age 9 months.

See page 232 (col. 1)  *See page 232 (col. 2)* 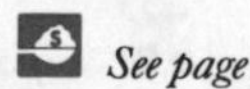 *See page 238*  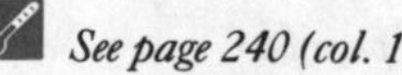 *See page 240 (col. 1)*  *See page 240 (col. 2)*

Chianti — California Style

READY: 1 YEAR

A full-bodied red wine to complement the popular new Cal-Mex cuisine.

PRIMARY INGREDIENTS

4 qt	California red grape concentrate	4 lit
1 lb	Dried wine grapes (chopped)	450 g
3 lb	Sugar	1.3 kg
6 tsp	Vinacid O	
6 qt	HOT water	6 lit
2 tsp	Yeast nutrient	
2 tsp	Pectic enzyme	
4 tsp	Liquid tannin	
4	Campden tablets (crushed)	
8 qt	COLD water	8 lit
1 pkt	Montpellier wine yeast	

SECONDARY INGREDIENTS

	Bentonite finings	
¼ tsp	Sulphite crystals	
1½ oz	Sinatin 17	45 ml

EQUIPMENT

Basic 10 + straining bag; coarse filter pads

PRIMARY SEQUENCE

1. Chop the grapes and place them, together with the grape concentrate, in the primary fermentor. Add the hot water, sugar, and Vinacid O.

2. Stir thoroughly until all the sugar is dissolved.

3. Add the next 5 ingredients. Mix well.

4. Check, and if necessary adjust, the specific gravity (SG) of the must. It should be 1.100.

5. Check, and if necessary adjust, the temperature of the must. It should be 75°F (23°C).

6. Add the yeast to a cup of warm water. Let stand for 10 minutes. Stir in.

7. Cover the fermentor with a plastic sheet; tie down. Keep in a warm place (75°F [23°C]). After 24 hours, check that fermentation has begun. Foam should be visible on the surface

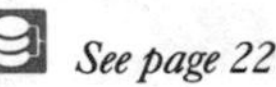

See page 220 · See page 223 · See page 224 · See page 226 · See page 227 · See page 229

and/or bubbles should be audible. If fermentation has not begun, see "Stuck Ferment," page 269.

8. Stir twice daily to keep the floating fruit moist.

9. Check SG every other day.

SECONDARY SEQUENCE

10. When SG reaches 1.020, scoop the grapes into a straining bag and squeeze it as dry as possible into the fermentor. Discard the pulp.

11. Rack into a clean carboy. Top up with cold tap water.

12. Attach the fermentation lock.

13. Move to a cooler location, ideally 65°F (18°C).

14. After 10 days or at SG 1.000, whichever comes first, rack into a clean carboy. Top up with cold tap water.

15. After 3 weeks or at SG .990–.995, whichever comes first, rack into a clean carboy.

16. Add the finings. Top up with cold tap water. Let rest 10 days.

17. Rack into the primary fermentor.

18. Filter into a clean carboy.

19. Add ¼ teaspoon of sulphite crystals dissolved in a small amount of water. Add the Sinatin 17. Top up with cold tap water.

20. Bulk age 3 months.

21. Bottle.

22. Bottle age 9 months.

See page 232 (col. 1)

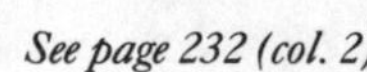
See page 232 (col. 2)

See page 238

See page 240 (col. 1)

See page 240 (col. 2)

Doradillo Wine

READY: 5 MONTHS

Grown in Australia, the Doradillo grape is often called Blanquette. It is a popular everyday table wine.

PRIMARY INGREDIENTS

3 qt	Doradillo white grape concentrate	3 lit
4 lb	Sugar	1.8 kg
14 tsp	Vinacid R	
6 qt	HOT water	6 lit
1 tsp	Yeast nutrient	
1 tsp	Pectic enzyme	
2 tsp	Liquid tannin	
9 qt	COLD water	9 lit
1 pkt	Champagne wine yeast	

SECONDARY INGREDIENTS

	Claro K. C. finings	
¼ tsp	Sulphite crystals	
8 oz	Wine conditioner	240 ml

EQUIPMENT

Basic 10; fine filter pads

PRIMARY SEQUENCE

1. Place the grape concentrate in the primary fermentor. Add the hot water, sugar, and Vinacid R.

2. Stir thoroughly until all the sugar is dissolved.

3. Add the next 4 ingredients. Mix well.

4. Check, and if necessary adjust, the specific gravity (SG) of the must. It should be 1.100.

5. Check, and if necessary adjust, the temperature of the must. It should be 75°F (23°C).

6. Add the yeast to a cup of warm water. Let stand for 10 minutes. Stir in.

7. Cover the fermentor with a plastic sheet; tie down. Keep in a warm place (75°F [23°C]). After 24 hours, check that fermentation has begun. Foam should be visible on the surface and/or bubbles should be audible. If

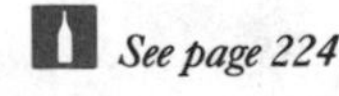

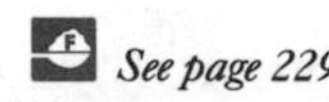

See page 220 See page 223 See page 224 See page 226 See page 227 See page 229

fermentation has not begun, see "Stuck Ferment," page 269.

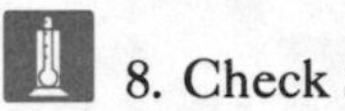

8. Check SG every other day.

SECONDARY SEQUENCE

9. When SG reaches 1.020, rack into a clean carboy. Top up with cold tap water.

10. Attach the fermentation lock.

11. Move to a cooler location, ideally 65°F (18°C).

12. After 10 days or at SG 1.000, whichever comes first, rack into a clean carboy. Top up with cold tap water.

13. After 3 weeks or at SG .990–.995, whichever comes first, rack into a clean carboy.

14. Add the finings. Top up with cold tap water. Let rest 10 days.

15. Rack into the primary fermentor.

16. Filter into a clean carboy.

17. Add ¼ teaspoon of sulphite crystals dissolved in a small amount of water. Top up with cold tap water.

18. Bulk age 1 month.

19. Add the wine conditioner and bottle.

20. Bottle age 3 months.

See page 232 (col. 1)  *See page 232 (col. 2)*  *See page 238*  *See page 240 (col. 1)* 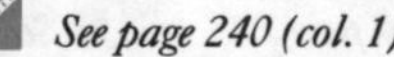  *See page 240 (col. 2)*

Fumé Blanc

READY: 1 YEAR

This can be a great wine. Made from the white Sauvignon grape, it has a crisp finish and a good bouquet. Deserves time in the cellar.

PRIMARY INGREDIENTS

3 qt	Sauvignon blanc grape concentrate	3 lit
2 lb	Fresh green table grapes (crushed)	1 kg
4 lb	Sugar	1.8 kg
2 tsp	Vinacid R	
6 qt	HOT water	6 lit
1 tsp	Yeast nutrient	
1 tsp	Pectic enzyme	
2 tsp	Liquid tannin	
8 qt	COLD water	8 lit
1 pkt	Narbonne wine yeast	

SECONDARY INGREDIENTS

	Claro K. C. finings	
¼ tsp	Sulphite crystals	

EQUIPMENT

Basic 10 + straining bag; fine filter pads

PRIMARY SEQUENCE

1. Crush the grapes and place them, together with the grape concentrate, in the primary fermentor. Add the hot water, sugar, and Vinacid R.

2. Stir thoroughly until all the sugar is dissolved.

3. Add the next 4 ingredients. Mix well.

4. Check, and if necessary adjust, the specific gravity (SG) of the must. It should be 1.100.

5. Check, and if necessary adjust, the temperature of the must. It should be 75°F (23°C).

6. Add the yeast to a cup of warm water. Let stand for 10 minutes. Stir in.

7. Cover the fermentor with a plastic sheet; tie down. Move to a warm place (75°F [23°C]). After 24 hours, check that fermentation has begun. Foam should be visible on the surface

See page 220   See page 223  See page 224  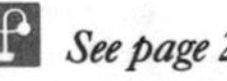 See page 226  See page 227  See page 229

and/or bubbles should be audible. If fermentation has not begun, see "Stuck Ferment," page 269.

8. Stir twice daily to keep the floating fruit moist.

9. Check SG every other day.

SECONDARY SEQUENCE

10. When SG reaches 1.020, scoop the grapes into a straining bag and squeeze it as dry as possible into the fermentor. Discard the pulp.

11. Rack into a clean carboy. Top up with cold tap water.

12. Attach the fermentation lock.

13. Move to a cooler location, ideally 65°F (18°C).

14. After 10 days or at SG 1.000, whichever comes first, rack into a clean carboy. Top up with cold tap water.

15. After 3 weeks or at SG .990–.995, whichever comes first, rack into a clean carboy.

16. Add the finings. Top up with cold tap water. Let rest 10 days.

17. Rack into the primary fermentor.

18. Filter into a clean carboy.

19. Add ¼ teaspoon of sulphite crystals dissolved in a small amount of water. Top up with cold tap water.

20. Bulk age 3 months.

21. Bottle.

22. Bottle age 9 months.

See page 232 (col. 1)

See page 232 (col. 2)

See page 238

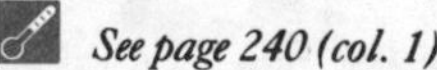
See page 240 (col. 1)

See page 240 (col. 2)

Gordo Muscat Dessert Wine

READY: 1 YEAR

A lusciously sweet wine from the Gordo grape, grown in Australia — not to be confused with the American Muscatel. A wine of indulgence for the classic dinner-party dessert.

PRIMARY INGREDIENTS		
3 qt	Gordo grape concentrate	3 lit
5½ lb	Sugar	2.5 kg
10 tsp	Vinacid R	
6 qt	HOT water	6 lit
2 tsp	Yeast nutrient	
1 tsp	Pectic enzyme	
2 tsp	Liquid tannin	
9 qt	COLD water	9 lit
1 pkt	Champagne wine yeast	

SECONDARY INGREDIENTS		
	Claro K. C. finings	
¼ tsp	Sulphite crystals	

EQUIPMENT
Basic 10; fine filter pads

PRIMARY SEQUENCE

1. Place the grape concentrate in the primary fermentor. Add the hot water, sugar, and Vinacid R.

2. Stir thoroughly until all the sugar is dissolved.

3. Add the next 4 ingredients. Mix well.

4. Check, and if necessary adjust, the specific gravity (SG) of the must. It should be 1.120.

5. Check, and if necessary adjust, the temperature of the must. It should be 75°F (23°C).

6. Add the yeast to a cup of warm water. Let stand for 10 minutes. Stir in.

7. Cover the fermentor with a plastic sheet; tie down. Keep in a warm place (75°F [23°C]). After 24 hours, check that fermentation has begun. Foam should be visible on the surface and/or bubbles should be audible. If fermen-

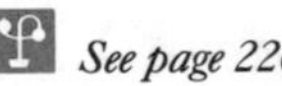

See page 220 See page 223 See page 224 See page 226 See page 227 See page 229

tation has not begun, see "Stuck Ferment," page 269.

8. Check SG every other day.

SECONDARY SEQUENCE

9. When SG reaches 1.020, rack into a clean carboy. Top up with cold tap water.

10. Attach the fermentation lock.

11. Move to a cooler location, ideally 65°F (18°C).

12. After 10 days, rack into a clean carboy. Top up with cold tap water.

13. After 3 weeks, rack into a clean carboy. Top up with cold tap water.

14. Allow the fermentation to cease; SG should be 1.000–1.005. Add the finings. Let rest 10 days.

15. Rack into a primary fermentor.

16. Filter into a clean carboy.

17. Add ¼ teaspoon of sulphite crystals dissolved in a small amount of water. Top up with cold tap water.

18. Bulk age 3 months.

19. Check the stability of the wine before bottling. (See "Residual Sugar," page 272.)

20. Bottle.

21. Bottle age 9 months.

See page 232 (col. 1)  *See page 232 (col. 2)*  *See page 238* 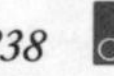  *See page 240 (col. 1)*  *See page 240 (col. 2)*

Graves

READY: 1 YEAR

Graves is a grape-growing area, between Médoc and Sauternes, on the west side of the Garonne River. Its Sauvignon blanc grapes produce a fruity white wine that ages well.

PRIMARY INGREDIENTS

3 qt	Australian Sauvignon blanc grape concentrate	3 lit
6 oz	Apple juice concentrate (frozen)	180 ml
4 lb	Sugar	1.8 kg
6 qt	HOT water	6 lit
2 tsp	Yeast nutrient	
2 tsp	Pectic enzyme	
2 tsp	Liquid tannin	
1 oz	Dried elderflowers (in small mesh bag)	30 g
8 qt	COLD water	8 lit
1 pkt	Cold-fermenting wine yeast	

SECONDARY INGREDIENTS

	Claro K. C. finings	
1 oz	Sinatin 17	30 ml
½ tsp	Sulphite crystals	
4 oz	Glycerine	120 ml

EQUIPMENT

Basic 10 + mesh bag for elderflowers; fine filter pads

PRIMARY SEQUENCE

1. Place the elderflowers in a small mesh bag.

2. Place the grape concentrate and the apple juice concentrate in the primary fermentor. Add the hot water and sugar.

3. Stir thoroughly until all the sugar is dissolved.

4. Add the next 5 ingredients. Mix well.

5. Test acid. Adjust, if necessary, to 5 g/lit.

See page 220

See page 223
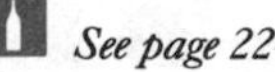
See page 224

See page 226

See page 227

See page 229

6. Check, and if necessary adjust, the specific gravity (SG) of the must. It should be 1.095.

7. Check, and if necessary adjust, the temperature of the must. It should be 70°F (21°C).

8. Add the yeast to a cup of warm water. Let stand for 10 minutes. Stir in.

9. Cover the fermentor with a plastic sheet; tie down. Keep in a warm place (70°F [21°C]). After 24 hours, check that fermentation has begun. Foam should be visible on the surface and/or bubbles should be audible. If fermentation has not begun, see "Stuck Ferment," page 269.

10. Stir twice daily for the first two days to keep the elderflowers moist, then remove the elderflowers.

11. Check SG every other day.

SECONDARY SEQUENCE

12. When SG reaches 1.020, rack into a clean carboy. Top up with cold tap water.

13. Attach the fermentation lock.

14. Move to a cooler location, ideally 60°F (15°C).

15. After 10 days or at SG 1.000, whichever comes first, rack into a clean carboy. Top up with cold tap water.

16. After 3 weeks or at SG .990–.995, whichever comes first, rack into a clean carboy.

17. Add the finings. Top up with cold tap water. Let rest 10 days.

18. Rack into the primary fermentor.

19. Filter into a clean carboy.

20. Add ¼ teaspoon of sulphite crystals dissolved in a small amount of water. Add the Sinatin 17. Top up with cold tap water.

21. Bulk age 2 months.

22. Rack into the primary fermentor. Add the glycerine and stir well.

23. Bottle.

24. Bottle age 10 months.

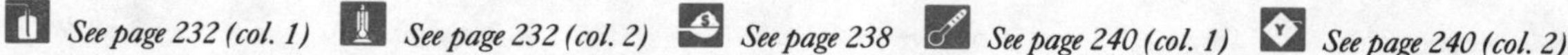

See page 232 (col. 1) *See page 232 (col. 2)* *See page 238* *See page 240 (col. 1)* *See page 240 (col. 2)*

Gewürztraminer White Table Wine

READY: 1 YEAR

'Traminer is the name given it nowadays; this is a German-style wine, made from one of the most aromatic grapes grown.

PRIMARY INGREDIENTS

3 qt	Gewürztraminer grape concentrate	3 lit
3 lb	Sugar	1.3 kg
6 tsp	Vinacid R	
6 qt	HOT water	6 lit
1 tsp	Yeast nutrient	
1 tsp	Pectic enzyme	
2 tsp	Liquid tannin	
9 qt	COLD water	9 lit
1 pkt	Champagne wine yeast	

SECONDARY INGREDIENTS

	Claro K. C. finings	
10 oz	White grape concentrate	300 ml
¼ tsp	Sulphite crystals	
2 tsp	Stabilizer	

EQUIPMENT

Basic 10; fine filter pads

PRIMARY SEQUENCE

1. Place the grape concentrate in the primary fermentor. Add the hot water, sugar, and Vinacid R.

2. Stir thoroughly until all the sugar is dissolved.

3. Add the next 4 ingredients. Mix well.

4. Check, and if necessary adjust, the specific gravity (SG) of the must. It should be 1.095.

5. Check, and if necessary adjust, the temperature of the must. It should be 75°F (23°C).

6. Add the yeast to a cup of warm water. Let stand for 10 minutes. Stir in.

7. Cover the fermentor with a plastic sheet; tie down. Keep in a warm place (75°F [23°C]). After 24 hours, check that fermentation has begun. Foam should be visible on the surface and/or bubbles should be audible. If

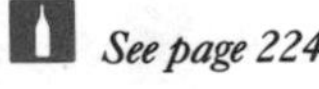

See page 220 See page 223 See page 224 See page 226 See page 227 See page 229

fermentation has not begun, see "Stuck Ferment," page 269.

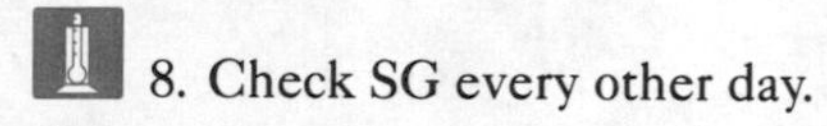

8. Check SG every other day.

SECONDARY SEQUENCE

9. When SG reaches 1.020, rack into a clean carboy. Top up with cold tap water.

10. Attach the fermentation lock.

11. Move to a cooler location, ideally 65°F (18°C).

12. After 10 days or at SG 1.000, whichever comes first, rack into a clean carboy. Top up with cold tap water.

13. After 3 weeks or at SG .990–.995, whichever comes first, rack into a clean carboy.

14. Add the finings. Top up with cold tap water. Let rest 10 days.

15. Rack into the primary fermentor. Add the white grape concentrate.

16. Filter into a clean carboy.

17. Stir in the stabilizer. Add ¼ teaspoon of sulphite crystals dissolved in a small amount of water. Top up with cold tap water.

18. Bulk age 2 months.

19. Bottle.

20. Bottle age 6 months.

See page 232 (col. 1)  *See page 232 (col. 2)*  *See page 238*  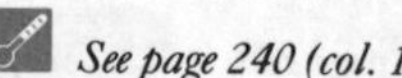 *See page 240 (col. 1)*  *See page 240 (col. 2)*

Johannisberg Riesling

READY: 7 MONTHS

The grape that produces famous German white wines

PRIMARY INGREDIENTS

3 qt	Johannisberg grape concentrate	3 lit
4 lb	Sugar	1.8 kg
6 tsp	Vinacid R	
6 qt	HOT water	6 lit
½ oz	Dried elderflowers (in small mesh bag)	15 g
3 tsp	Liquid tannin	
2 tsp	Pectic enzyme	
2 tsp	Yeast nutrient	
9 qt	COLD water	9 lit
1 pkt	Champagne wine yeast	

SECONDARY INGREDIENTS

1 oz	Sinatin 17	30 ml
9 oz	Wine conditioner	300 ml
	Claro K. C. finings	

EQUIPMENT

Basic 10 + mesh bag for elderflowers; fine filter pads

PRIMARY SEQUENCE

1. Place the elderflowers in a small mesh bag.

2. Place the grape concentrate in the primary fermentor. Add the hot water, sugar, and Vinacid R.

3. Stir thoroughly until all the sugar is dissolved.

4. Add the next 5 ingredients. Mix well.

5. Check, and if necessary adjust, the temperature of the must. It should be 75°F (23°C).

6. Add the yeast to a cup of warm water. Let stand for 10 minutes. Stir in.

7. Cover the fermentor with a plastic sheet; tie down. Keep in a warm place (75°F [23°C]). After 24 hours, check that fermentation has begun. Foam should be visible on the surface and/or bubbles should be audible. If fermen-

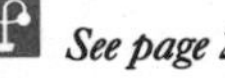

See page 220 See page 223 See page 224 See page 226 See page 227 See page 229

tation has not begun, see "Stuck Ferment," page 269.

8. Stir twice daily to keep the elderflowers moist.

9. Check the specific gravity (SG) every other day.

SECONDARY SEQUENCE

10. When SG reaches 1.020, remove the mesh bag of elderflowers and rack into a clean carboy. Top up with cold tap water.

11. Attach the fermentation lock.

12. Move to a cooler location, ideally 65°F (18°C).

13. After 10 days or at SG 1.000, whichever comes first, rack into a clean carboy. Top up with cold tap water.

14. After 3 weeks, rack into a clean carboy. Top up with cold tap water.

15. At SG .990–.995, add the finings. Let rest 10 days.

16. Rack into the primary fermentor.

17. Filter into a clean carboy.

18. Add ¼ teaspoon of sulphite crystals dissolved in a small amount of water. Add the Sinatin 17 and the wine conditioner.

19. Bulk age 1 month.

20. Bottle.

21. Bottle age 5 months.

See page 232 (col. 1) *See page 232 (col. 2)* *See page 238* *See page 240 (col. 1)* *See page 240 (col. 2)*

Johannisberg Riesling (Spätlese)

READY: 7 MONTHS

A German wine from late-harvested grapes, full-bodied and sweet

PRIMARY INGREDIENTS

3 qt	Johannisberg grape concentrate	3 lit
9 qt	HOT water	9 lit
4 lb	Sugar	1.8 kg
5 tsp	Vinacid R	
2 lb	Fresh green grapes (crushed)	1 kg
2 tsp	Liquid tannin	
2 tsp	Yeast nutrient	
2 tsp	Pectic enzyme	
½ oz	Dried elderflowers (in a mesh bag)	15 g
5 qt	COLD water	5 lit
1 pkt	Champagne wine yeast	

SECONDARY INGREDIENTS

10 oz	Wine conditioner	300 ml
	Claro K. C. finings	

EQUIPMENT

Basic 10 + straining bag, small mesh bag for elderflowers, and potato masher; fine filter pads

PRIMARY SEQUENCE

1. Place the elderflowers in a small mesh bag. Crush the grapes.

2. Place the grape concentrate in the primary fermentor. Add the hot water, sugar, and Vinacid R.

3. Stir thoroughly until all the sugar is dissolved.

4. Add the next 6 ingredients. Mix well.

5. Check, and if necessary adjust, the specific gravity (SG) of the must. It should be 1.100.

6. Check, and if necessary adjust, the temperature of the must. It should be 75°F (23°C).

7. Add the yeast to a cup of warm water. Let stand for 10 minutes. Stir in.

8. Cover the fermentor with a plastic sheet; tie down. Keep in a warm place (75°F [23°C]).

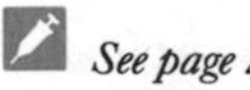

See page 220 See page 223 See page 224 See page 226 See page 227 See page 229

After 24 hours, check that fermentation has begun. Foam should be visible on the surface and/or bubbles should be audible. If fermentation has not begun, see "Stuck Ferment," page 269.

9. Stir twice daily to keep the floating grapes and elderflowers moist.

10. After 4 days remove the mesh bag of elderflowers. Scoop the grapes into a straining bag and squeeze the juice gently into the fermentor. Discard the pulp.

11. Check SG every other day.

SECONDARY SEQUENCE

12. When SG reaches 1.020, rack into a clean carboy. Top up with cold tap water.

13. Attach the fermentation lock.

14. Move to a cooler location, ideally 65°F (18°C).

15. After 10 days or at SG 1.000, whichever comes first, rack into a clean carboy. Top up with cold tap water.

16. After 3 weeks or at SG .990–.995, whichever comes first, rack into a clean carboy.

17. Add the finings. Top up with cold tap water. Let rest 10 days.

18. Rack into the primary fermentor.

19. Filter into a clean carboy.

20. Bulk age 1 month.

21. Add the wine conditioner and bottle.

22. Bottle age 5 months.

 *See page 232 (col. 1)* 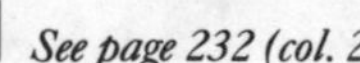 *See page 232 (col. 2)*  *See page 238*  *See page 240 (col. 1)* 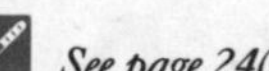  *See page 240 (col. 2)*

Moselle

READY: 7 MONTHS

A soft, German-style white wine; rich, fragrant, and slightly sweet

PRIMARY INGREDIENTS

Imperial	Ingredient	Metric
3 qt	Australian Gordo grape concentrate	3 lit
1 qt	Doradillo grape concentrate	1 lit
1 lb	Clover honey	450 g
1 lb	Sugar	450 g
12 tsp	Vinacid R	
6 qt	HOT water	6 lit
2 tsp	Yeast nutrient	
2 tsp	Pectic enzyme	
2 tsp	Liquid tannin	
1 oz	Dried elderflowers (in small mesh bag)	30 g
8 qt	COLD water	8 lit
1 pkt	Champagne wine yeast	

SECONDARY INGREDIENTS

Imperial	Ingredient	Metric
	Claro K. C. finings	
¼ tsp	Sulphite crystals	
10 oz	Wine conditioner	300 ml

EQUIPMENT

Basic 10 + straining bag, mesh bag for elderflowers; fine filter pads

PRIMARY SEQUENCE

1. Place the elderflowers in a small mesh bag.
2. Place the grape concentrates and honey in the primary fermentor.
3. Add the hot water, sugar, and Vinacid R.
4. Stir thoroughly until all the sugar is dissolved.
5. Add the next 5 ingredients. Mix well.
6. Check, and if necessary adjust, the specific gravity (SG) of the must. It should be 1.090.

See page 220  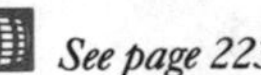 See page 223 See page 224   See page 226  See page 227  See page 229

7. Check, and if necessary adjust, the temperature of the must. It should be 75°F (23°C).

8. Add the yeast to a cup of warm water. Let stand for 10 minutes. Stir in.

9. Cover the fermentor with a plastic sheet; tie down. Keep in a warm place (75°F [23°C]). After 24 hours, check that fermentation has begun. Foam should be visible on the surface and/or bubbles should be audible. If fermentation has not begun, see "Stuck Ferment," page 269.

10. Stir twice daily to keep the elderflowers moist.

11. Check SG every other day.

SECONDARY SEQUENCE

12. When SG reaches 1.020, remove the mesh bag of elderflowers and rack into a clean carboy. Top up with cold tap water.

13. Attach the fermentation lock.

14. Move to a cooler location, ideally 65°F (18°C).

15. After 10 days or at SG 1.000, whichever comes first, rack into a clean carboy. Top up with cold tap water.

16. After 3 weeks or at SG .990–.995, whichever comes first, rack into a clean carboy.

17. Add the finings. Top up with cold tap water. Let rest 10 days.

18. Rack into the primary fermentor.

19. Filter into a clean carboy.

20. Add ¼ teaspoon of sulphite crystals dissolved in a small amount of water. Top up with cold tap water.

21. Bulk age 1 month.

22. Rack into the primary fermentor.

23. Add the wine conditioner and stir in.

24. Bottle.

25. Bottle age 5 months.

See page 232 (col. 1) 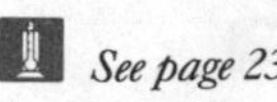 *See page 232 (col. 2)*  *See page 238*  *See page 240 (col. 1)* 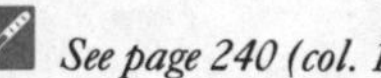  *See page 240 (col. 2)*

Pedro Ximenez

READY: 7 MONTHS

Pedro Ximenez grapes grown in Australia differ from those grown in Spain for use in sherries. This recipe produces a table wine, pale in color, fresh tasting and early maturing.

PRIMARY INGREDIENTS

3 qt	Pedro Ximenez grape concentrate	3 lit
4 lb	Sugar	1.8 kg
10 tsp	Vinacid R	
6 qt	HOT water	6 lit
1 tsp	Yeast nutrient	
2 tsp	Pectic enzyme	
2 tsp	Liquid tannin	
9 qt	COLD water	9 lit
1 pkt	Cold-fermenting wine yeast	

SECONDARY INGREDIENTS

	Claro K. C. finings	
¼ tsp	Sulphite crystals	
8 oz	Wine conditioner	240 ml

EQUIPMENT

Basic 10; fine filter pads

PRIMARY SEQUENCE

1. Place the grape concentrate in the primary fermentor. Add the hot water, sugar, and Vinacid R.

2. Stir thoroughly until all the sugar is dissolved.

3. Add the next 4 ingredients. Mix well.

4. Check, and if necessary adjust, the specific gravity (SG) of the must. It should be 1.095.

5. Check, and if necessary adjust, the temperature of the must. It should be 70°F (21°C).

6. Add the yeast to a cup of warm water. Let stand for 10 minutes. Stir in.

7. Cover the fermentor with a plastic sheet; tie down. Keep in a warm place (70°F [21°C]). After 24 hours, check that fermentation has begun. Foam should be visible on the surface and/or bubbles should be audible. If

See page 220  See page 223  See page 224 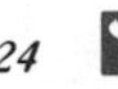 See page 226  See page 227 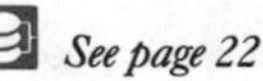  See page 229

fermentation has not begun, see "Stuck Ferment," page 269.

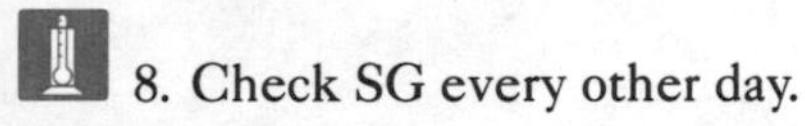

8. Check SG every other day.

SECONDARY SEQUENCE

9. When SG reaches 1.020, rack into a clean carboy. Top up with cold tap water.

10. Attach the fermentation lock.

11. Move to a cooler location, ideally 60°F (15°C).

12. After 10 days or at SG 1.000, whichever comes first, rack into a clean carboy. Top up with cold tap water.

13. After 3 weeks or at SG .990–.995, whichever comes first, rack into a clean carboy.

14. Add the finings. Top up with cold tap water. Let rest 10 days.

15. Rack into the primary fermentor.

16. Filter into a clean carboy.

17. Add ¼ teaspoon of sulphite crystals dissolved in a small amount of water. Top up with cold tap water.

18. Bulk age 1 month.

19. Add the wine conditioner and bottle.

20. Bottle age 5 months.

See page 232 (col. 1)  *See page 232 (col. 2)* 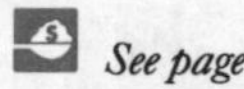 *See page 238*  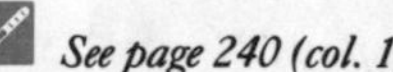 *See page 240 (col. 1)*  *See page 240 (col. 2)*

Prize-Winning Dry Red Table Wine

READY: 2 YEARS

1st-prize winner in a 1986 competition

PRIMARY INGREDIENTS

3 qt	Australian Grenache concentrate	3 lit
2¼ qt	Spanish red concentrate	2.3 lit
6 qt	HOT water	6 lit
2 tsp	Liquid tannin	
8 qt	COLD water	8 lit
1 pkt	Montpellier wine yeast	

SECONDARY INGREDIENTS

	Gelatin finings	
2 oz	Oak chips or sticks	60 g
¼ tsp	Sulphite crystals	

EQUIPMENT

Basic 10; coarse filter pads

PRIMARY SEQUENCE

1. Place the grape juice concentrates in the primary fermentor. Add the hot water.

2. Stir thoroughly.

3. Add the next 2 ingredients. Mix well.

4. Check, and if necessary adjust, the specific gravity (SG) of the must. It should be 1.100.

5. Check, and if necessary adjust, the temperature of the must. It should be 75°F (23°C).

6. Add the yeast to a cup of warm water. Let stand for 10 minutes. Stir in.

7. Cover the fermentor with a plastic sheet; tie down. Keep in a warm place (75°F [23°C]). After 24 hours, check that fermentation has begun. Foam should be visible on the surface and/or bubbles should be audible. If fermentation has not begun, see "Stuck Ferment," page 269.

8. Check SG every other day.

 See page 220 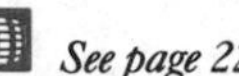 See page 223  See page 224  See page 226  See page 227  See page 229

SECONDARY SEQUENCE

9. When SG reaches 1.020, rack into a clean carboy. Top up with cold tap water.

10. Attach the fermentation lock.

11. Move to a cooler location, ideally 65°F (18°C).

12. After 10 days or at SG 1.000, whichever comes first, rack into a clean carboy. Top up with cold tap water.

13. After 3 weeks, rack into a clean carboy into which you have placed the oak chips. Top up with cold tap water. Let rest 1 month.

14. Add the finings and let rest 10 days.

15. Rack into the primary fermentor.

16. Filter into a clean carboy.

17. Add ¼ teaspoon of sulphite crystals dissolved in a small amount of water. Top up with cold tap water.

18. Bulk age 3 months.

19. Bottle.

20. Bottle age 18 months.

See page 232 (col. 1)  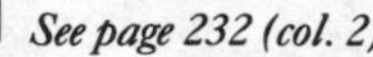 *See page 232 (col. 2)*  *See page 238*  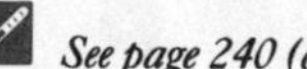 *See page 240 (col. 1)*  *See page 240 (col. 2)*

Rhône Red

READY: 2 YEARS

This is a full-bodied red wine that will keep well and can exceed 13% alcohol by volume. Rich, earthy, robust.

PRIMARY INGREDIENTS

3 qt	Australian Shiraz concentrate	3 lit
2¼ qt	Spanish red grape concentrate	2.3 lit
1 lb	Dried Shiraz wine grapes (chopped)	400 g
6 tsp	Vinacid O	
6 qt	HOT water	6 lit
2 tsp	Yeast nutrient	
2 tsp	Pectic enzyme	
2 tsp	Liquid tannin	
8 qt	COLD water	8 lit
1 pkt	Montpellier wine yeast	

SECONDARY INGREDIENTS

	Gelatin finings	
¼ tsp	Sulphite crystals	
1½ oz	Sinatin 17	45 ml

EQUIPMENT

Basic 10 + straining bag; coarse filter pads

PRIMARY SEQUENCE

1. Chop the dried grapes and place them, together with the grape concentrate, in the primary fermentor. Add the hot water and Vinacid O.

2. Stir thoroughly until all the sugar is dissolved.

3. Add the next 4 ingredients. Mix well.

4. Check, and if necessary adjust, the specific gravity (SG) of the must. It should be 1.110.

5. Check, and if necessary adjust, the temperature of the must. It should be 75°F (23°C).

6. Add the yeast to a cup of warm water. Let stand for 10 minutes. Stir in.

7. Cover the fermentor with a plastic sheet; tie down. Keep in a warm place (75°F [23°C]). After 24 hours, check that fermentation has begun. Foam should be visible on the surface

See page 220 See page 223 See page 224 See page 226 See page 227 See page 229

and/or bubbles should be audible. If fermentation has not begun, see "Stuck Ferment," page 269.

8. Stir twice daily to keep the floating fruit moist.

9. Check SG every other day.

SECONDARY SEQUENCE

10. When SG reaches 1.020, scoop out the grapes into a straining bag and squeeze as dry as possible into the fermentor. Discard the pulp.

11. Rack into a clean carboy. Top up with cold tap water.

12. Attach the fermentation lock.

13. Move to a cooler location, ideally 65°F (18°C).

14. After 10 days or at SG 1.000, whichever comes first, rack into a clean carboy. Top up with cold tap water.

15. After 3 weeks or at SG .990–.995, whichever comes first, rack into a clean carboy.

16. Add the finings. Top up with cold tap water. Let rest 10 days.

17. Rack into the primary fermentor.

18. Filter into a clean carboy.

19. Add ¼ teaspoon of sulphite crystals dissolved in a small amount of water. Add the Sinatin 17. Top up with cold tap water.

20. Bulk age 3 months.

21. Rack and top up.

22. Bulk age again 3 months.

23. Bottle.

24. Bottle age 18 months.

See page 232 (col. 1)   *See page 232 (col. 2)*  *See page 238*  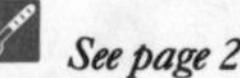 *See page 240 (col. 1)*  *See page 240 (col. 2)*

Ruby Port

READY: 18 MONTHS

Use the extra concentrate called for, and you will have a true port-style wine.

PRIMARY INGREDIENTS

4 qt	Shiraz grape concentrate	4 lit
2 lb	Glucose solids	1 kg
5 oz	Dried elderberries (in a small mesh bag)	150 g
4 tsp	Vinacid O	
6 qt	HOT water	6 lit
2 tsp	Yeast nutrient	
2 tsp	Pectic enzyme	
2 tsp	Liquid tannin	
7 qt	COLD water	7 lit
1 pkt	Wine yeast with a high alcohol tolerance	

SECONDARY INGREDIENTS

1 qt	Sugar syrup (See "Syrup Feeding," page 274.)	1 lit

(continues)

	Gelatin finings	
1 oz	Sinatin 17	30 ml
26 oz	Vodka	780 ml
¼ tsp	Sulphite crystals	
10 oz	Wine conditioner	300 ml

EQUIPMENT

Basic 10 + small mesh bag; coarse filter pads

PRIMARY SEQUENCE

1. Put the dried elderberries in a small mesh bag and place them, together with the grape concentrate and glucose, in the primary fermentor. Add the hot water and Vinacid O.

2. Stir thoroughly until all the sugar is dissolved.

3. Add the next 4 ingredients. Mix well.

4. Check, and if necessary adjust, the specific gravity (SG) of the must. It should be 1.100.

5. Check, and if necessary adjust, the temperature of the must. It should be 75°F (23°C).

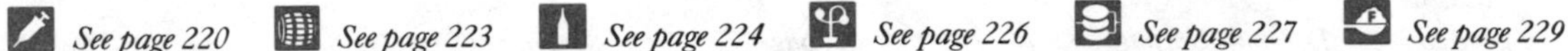

 *See page 220*  *See page 223* *See page 224* *See page 226*  *See page 227*  *See page 229*

6. Add the yeast to a cup of warm water. Let stand for 10 minutes. Stir in.

7. Cover the fermentor with a plastic sheet; tie down. Keep in a warm place (75°F [23°C]). After 24 hours, check that fermentation has begun. Foam should be visible on the surface and/or bubbles should be audible. If fermentation has not begun, see "Stuck Ferment," page 269.

8. Stir twice daily to keep the floating elderberries moist.

9. Check SG every other day.

SECONDARY SEQUENCE

10. When SG reaches 1.020, remove the mesh bag of elderberries and squeeze it dry into the fermentor. Discard.

11. Rack into a clean carboy. Top up with cold tap water.

12. Attach the fermentation lock and continue to ferment in a warm place.

13. When SG falls to 1.000, add 2 cups of sugar syrup. Continue to ferment in a warm place.

14. If SG falls below 1.000 again, add the remainder of the sugar syrup. If SG does not fall below 1.000 again, do not add the remainder of the syrup, but allow to ferment out.

15. When fermentation ceases, rack into a clean carboy. Top up with cold tap water.

16. After 4 weeks, add the finings and let rest 10 days.

17. Rack into the primary fermentor.

18. Filter into a clean carboy.

19. Add ¼ teaspoon of sulphite crystals dissolved in a small amount of water. Add the Sinatin 17. Top up with cold tap water.

20. Bulk age 3 months.

21. Rack into the primary fermentor. Add the vodka and wine conditioner. (See "Fortifying Wines with Distilled Alcohol," page 255.)

22. Bottle.

23. Bottle age 1 year.

See page 232 (col. 1)  *See page 232 (col. 2)* 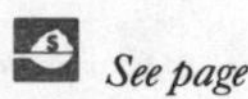 *See page 238*  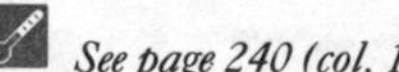 *See page 240 (col. 1)*  *See page 240 (col. 2)*

Sauterne

READY: 1 YEAR

A medium-class, full-bodied wine, lusciously sweet; made from the noble Sauvignon blanc grape

PRIMARY INGREDIENTS

3 qt	Sauvignon blanc grape concentrate	3 lit
4 lb	Sugar	1.8 kg
2 tsp	Vinacid O	
6 qt	HOT water	6 lit
2 tsp	Yeast nutrient	
1 tsp	Pectic enzyme	
2 tsp	Liquid tannin	
9 qt	COLD water	9 lit
1 pkt	Champagne wine yeast	

SECONDARY INGREDIENTS

	Claro K. C. finings	
6 oz	Frozen apple juice concentrate	180 ml
½ tsp	Sulphite crystals	
8 oz	Wine conditioner	240 ml

EQUIPMENT

Basic 10; fine filter pads

PRIMARY SEQUENCE

1. Place the grape concentrate in the primary fermentor. Add the hot water, sugar, and Vinacid O.

2. Stir thoroughly until all the sugar is dissolved.

3. Add the next 4 ingredients. Mix well.

4. Check, and if necessary adjust, the specific gravity (SG) of the must. It should be 1.110.

5. Check, and if necessary adjust, the temperature of the must. It should be 75°F (23°C).

6 Add the yeast to a cup of warm water. Let stand for 10 minutes. Stir in.

7. Cover the fermentor with a plastic sheet; tie down. Keep in a warm place (75°F [23°C]). After 24 hours, check that fermentation has begun. Foam should be visible on the surface and/or bubbles should be audible. If

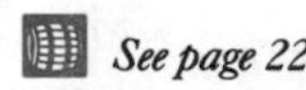

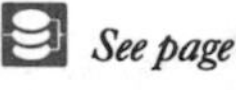
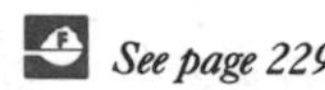

See page 220 *See page 223* *See page 224* *See page 226* *See page 227* *See page 229*

fermentation has not begun, see "Stuck Ferment," page 269.

8. Check SG every other day.

SECONDARY SEQUENCE

9. When SG reaches 1.020, rack into a clean carboy. Top up with cold tap water.

10. Attach the fermentation lock.

11. Move to a cooler location, ideally 65°F (18°C).

12. After 10 days or at SG 1.000, whichever comes first, rack into a clean carboy. Top up with cold tap water.

13. After 3 weeks or at SG .990–.995, whichever comes first, rack into a clean carboy.

14. Add the finings. Top up with cold tap water. Let rest 10 days.

15. Rack into the primary fermentor and add the apple juice concentrate.

16. Filter into a clean carboy.

17. Add the wine conditioner and ½ teaspoon of sulphite crystals dissolved in a small amount of water. Top up with cold tap water.

18. Bulk age 3 months.

19. Bottle.

20. Bottle age 9 months.

See page 232 (col. 1) *See page 232 (col. 2)* *See page 238* *See page 240 (col. 1)* *See page 240 (col. 2)*

Shiraz (Sirah) Red Table Wine

READY: 1 YEAR

Shiraz is Australia's staple red wine grape. It is related to the Sirah of the north Rhône and is often called Hermitage. It makes a superb wine.

PRIMARY INGREDIENTS

3 qt	Australian Shiraz grape concentrate	3 lit
3¼ lb	Sugar	1.5 kg
4 tsp	Vinacid O	
6 qt	HOT water	6 lit
1 lb	Dried Shiraz grapes (coarsely chopped)	450 g
1 tsp	Yeast nutrient	
1 tsp	Pectic enzyme	
2 tsp	Liquid tannin	
9 qt	COLD water	9 lit
1 pkt	Montpellier wine yeast	

SECONDARY INGREDIENTS

	Claro K. C. finings	
1 oz	Sinatin 17	30 ml
¼ tsp	Sulphite crystals	

(continues)

EQUIPMENT

Basic 10 + mesh bag; coarse filter pads

PRIMARY SEQUENCE

1. Place the grape concentrate in the primary fermentor. Add the hot water, sugar, and Vinacid O.

2. Stir thoroughly until all the sugar is dissolved.

3. Coarsely chop the dried grapes and add them to the fermentor. For convenience, they may be placed in a mesh bag.

4. Add the next 4 ingredients. Mix well.

5. Check, and if necessary adjust, the temperature of the must. It should be 75°F (23°C).

6. Add the yeast to a cup of warm water. Let stand for 10 minutes. Stir in.

7. Cover the fermentor with a plastic sheet; tie down. Keep in a warm place (75°F [23°C]).

See page 220   See page 223  See page 224  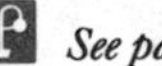 See page 226  See page 227  See page 229

After 24 hours, check that fermentation has begun. Foam should be visible on the surface and/or bubbles should be audible. If fermentation has not begun, see "Stuck Ferment," page 269.

8. Stir twice daily to keep the floating fruit moist.

9. Check the specific gravity (SG) every other day.

SECONDARY SEQUENCE

10. When SG reaches 1.020, remove the chopped grapes and rack into a clean carboy. Top up with cold tap water.

11. Attach the fermentation lock.

12. Move to a cooler location, ideally 65°F (18°C).

13. After 10 days or at SG 1.000, whichever comes first, rack into a clean carboy. Top up with cold tap water.

14. After 3 weeks or at SG .990–.995, whichever comes first, rack into a clean carboy.

15. Add the finings. Top up with cold tap water. Let rest 10 days.

16. Rack into the primary fermentor.

17. Filter into a clean carboy.

18. Add ¼ teaspoon of sulphite crystals dissolved in a small amount of water. Add the Sinatin 17. Top up with cold tap water.

19. Bulk age 3 months.

20. Bottle.

21. Bottle age 9 months.

See page 232 (col. 1)  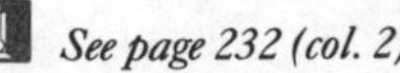 *See page 232 (col. 2)*  *See page 238*  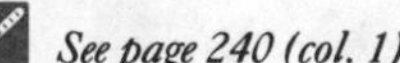 *See page 240 (col. 1)*  *See page 240 (col. 2)*

Spanish Red Table Wine

READY: 1 YEAR

Medium-bodied dry house wine

PRIMARY INGREDIENTS

2¼ qt	Spanish red grape concentrate	2.3 lit
4½ lb	Sugar	2 kg
7 tsp	Vinacid O	
6 qt	HOT water	6 lit
2 tsp	Yeast nutrient	
2 tsp	Liquid tannin	
9 qt	COLD water	9 lit
1 pkt	Montpellier wine yeast	

SECONDARY INGREDIENTS

	Bentonite finings	
¼ tsp	Sulphite crystals	

EQUIPMENT

Basic 10; coarse filter pads

PRIMARY SEQUENCE

1. Place the grape concentrate in the primary fermentor. Add the hot water, sugar, and Vinacid O.

2. Stir thoroughly until all the sugar is dissolved.

3. Add the next 3 ingredients. Mix well.

4. Check, and if necessary adjust, the specific gravity (SG) of the must. It should be 1.100.

5. Check, and if necessary adjust, the temperature of the must. It should be 75°F (23°C).

6. Add the yeast to a cup of warm water. Let stand for 10 minutes. Stir in.

7. Cover the fermentor with a plastic sheet; tie down. Keep in a warm place (75°F [23°C]). After 24 hours, check that fermentation has begun. Foam should be visible on the surface and/or bubbles should be audible. If fermentation has not begun, see "Stuck Ferment," page 269.

8. Check the SG every other day.

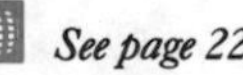

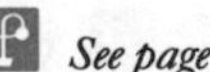

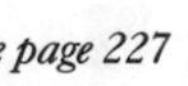
See page 220 See page 223 See page 224 See page 226 See page 227 See page 229

SECONDARY SEQUENCE

9. When SG reaches 1.020, rack into a clean carboy. Top up with cold tap water.

10. Attach the fermentation lock.

11. Move to a cooler location, ideally 65°F (18°C).

12. After 10 days or at SG 1.000, whichever comes first, rack into a clean carboy. Top up with cold tap water.

13. After 3 weeks or at SG .990–.995, whichever comes first, rack into a clean carboy.

14. Add the finings. Top up with cold tap water. Let rest 10 days.

15. Rack into the primary fermentor.

16. Filter into a clean carboy.

17. Add ¼ teaspoon of sulphite crystals dissolved in a small amount of water. Top up with cold tap water.

18. Bulk age 2 months.

19. Bottle.

20. Bottle age 10 months.

See page 232 (col. 1)  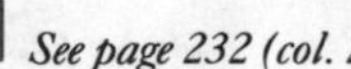 *See page 232 (col. 2)*  *See page 238*  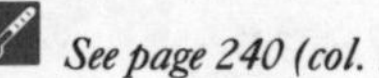 *See page 240 (col. 1)*  *See page 240 (col. 2)*

Spanish White Table Wine

READY: 1 YEAR

Reminiscent of white Burgundy, but without the bouquet; has an excellent vinous quality

PRIMARY INGREDIENTS

2¼ qt	Spanish white grape concentrate	2.3 lit
4 lb	Sugar	1.8 kg
6 tsp	Vinacid O	
6 qt	HOT water	6 lit
2 tsp	Yeast nutrient	
2 tsp	Liquid tannin	
9 qt	COLD water	9 lit
1 pkt	Champagne wine yeast	

SECONDARY INGREDIENTS

	Bentonite finings	
¼ tsp	Sulphite crystals	

EQUIPMENT

Basic 10; fine filter pads

PRIMARY SEQUENCE

1. Place the grape concentrate in the primary fermentor. Add the hot water, sugar, and Vinacid O.

2. Stir thoroughly until all the sugar is dissolved.

3. Add the next 3 ingredients. Mix well.

4. Check, and if necessary adjust, the specific gravity (SG) of the must. It should be 1.095.

5. Check, and if necessary adjust, the temperature of the must. It should be 75°F (23°C).

6. Add the yeast to a cup of warm water. Let stand for 10 minutes. Stir in.

7. Cover the fermentor with a plastic sheet; tie down. Keep in a warm place (75°F [23°C]). After 24 hours, check that fermentation has begun. Foam should be visible on the surface and/or bubbles should be audible. If fermentation has not begun, see "Stuck Ferment," page 269.

8. Check the SG every other day.

See page 220

See page 223

See page 224

See page 226

See page 227

See page 229

SECONDARY SEQUENCE

9. When SG reaches 1.020, rack into a clean carboy. Top up with cold tap water.

10. Attach the fermentation lock.

11. Move to a cooler location, ideally 65°F (18°C).

12. After 10 days or at SG 1.000, whichever comes first, rack into a clean carboy. Top up with cold tap water.

13. After 3 weeks or at SG .990–.995, whichever comes first, rack into a clean carboy.

14. Add the finings. Top up with cold tap water. Let rest 10 days.

15. Rack into the primary fermentor.

16. Filter into a clean carboy.

17. Add ¼ teaspoon of sulphite crystals dissolved in a small amount of water. Top up with cold tap water.

18. Bulk age 2 months.

19. Bottle.

20. Bottle age 10 months.

See page 232 (col. 1)  *See page 232 (col. 2)* 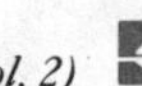 *See page 238*  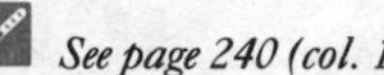 *See page 240 (col. 1)*  *See page 240 (col. 2)*

Valpolicella

READY: 1 YEAR

From the heart of Italy's wine-growing Veneto district, Valpolicella is a robust red, full bodied and flavorful.

PRIMARY INGREDIENTS

4¼ qt	Valpolicella Italian grape concentrate	4.3 lit
5 oz	Dried elderberries (in a small mesh bag)	150 g
3¼ lb	Sugar	1.5 kg
3 tsp	Vinacid O	
6 qt	HOT water	6 lit
1 tsp	Yeast nutrient	
1 tsp	Liquid tannin	
8 qt	COLD water	8 lit
1 pkt	Montpellier wine yeast	

SECONDARY INGREDIENTS

	Isinglass finings	
¼ tsp	Sulphite crystals	

EQUIPMENT

Basic 10 + mesh bag for elderberries; coarse filter pads

PRIMARY SEQUENCE

1. Put the elderberries in a small mesh bag and place them, together with the grape concentrate, in the primary fermentor. Add the hot water, sugar, and Vinacid O.

2. Stir thoroughly until all the sugar is dissolved.

3. Add the next 3 ingredients. Mix well.

4. Check, and if necessary adjust, the specific gravity (SG) of the must. It should be 1.100.

5. Check, and if necessary adjust, the temperature of the must. It should be 75°F (23°C).

6. Add the yeast to a cup of warm water. Let stand for 10 minutes. Stir in.

7. Cover the fermentor with a plastic sheet; tie down. Keep in a warm place (75°F [23°C]). After 24 hours, check that fermentation has

See page 220

See page 223

See page 224

See page 226

See page 227
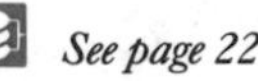

See page 229

begun. Foam should be visible on the surface and/or bubbles should be audible. If fermentation has not begun, see "Stuck Ferment," page 269.

8. Stir twice daily to keep the floating elderberries moist.

9. Check SG every other day.

SECONDARY SEQUENCE

10. When SG reaches 1.020, remove the mesh bag of elderberries. Discard.

11. Rack into a clean carboy. Top up with cold tap water.

12. Attach the fermentation lock.

13. Move to a cooler location, ideally 65°F (18°C).

14. After 10 days or at SG 1.000, whichever comes first, rack into a clean carboy. Top up with cold tap water.

15. After 3 weeks or at SG .990–.995, whichever comes first, rack into a clean carboy.

16. Add the finings. Top up with cold tap water. Let rest 10 days.

17. Rack into the primary fermentor.

18. Filter into a clean carboy.

19. Add ¼ teaspoon of sulphite crystals dissolved in a small amount of water. Top up with cold tap water.

20. Bulk age 3 months.

21. Bottle.

22. Bottle age 9 months.

See page 232 (col. 1)  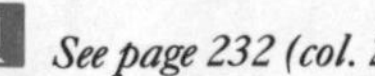 *See page 232 (col. 2)*  *See page 238*  *See page 240 (col. 1)*  *See page 240 (col. 2)*

FRESH GRAPE WINES

QUICK REFERENCE FOR FRESH GRAPE WINES

Before You Begin

1. Select grapes appropriate to your needs. (See "Purchasing Fresh Grapes," page 263.)

2. Confirm the date of delivery from the importer. Check whether the service includes crushing.

3. If renting equipment, make reservations. You will need the Basic 10, plus an 8-foot (2.5-meter) syphon hose, an acid-testing kit, a grape crusher, a grape press, and a grape destemmer. (See "Crushing and Pressing Grapes," page 249.)

4. Calculate the containers you will need.

1 lug of grapes equals 35 lb (15.9 kg) and makes 2 gallons (10 bottles).

12 lugs of grapes equals 420 lb (190.9 kg) and makes 24 gallons (120 bottles).

For 12 lugs you will need a 45-gallon primary fermentor.

5. If using a 45-gallon fermentor, decide beforehand where to put it. It should be 3 feet (1 m) off the floor, with room to set a grape crusher on top. The ambient temperature should be 70°F (21°C).

6. If you intend to use barrels for secondary fermentation, clean and prepare them for use. (See "Barrels," page 244.)

7. Decide whether or not you will make a second run. Note that white grapes are pressed within 24 hours of crushing. This means that you will be starting your second run while your first run is still in the primary fermentor. You will therefore need an additional primary fermentor.

8. Clean all equipment and containers before use with sanitizing solution.

Topping Up Fresh Grape Wines

Because of the large quantity of lees produced by fresh grape wines, you will need to add large volumes of liquid after each racking.

Tap water in volumes larger than 4 ounces per gallon (30 ml per liter) will impair the quality of the wine. It will therefore be necessary to keep a quantity of top-up wine on hand. You can keep the risks of top-up containers to a minimum if you follow a few simple precautions:

1. Never keep top-up wine in quantities of less than ½ gallon.

2. Keep top-up wine in full containers only.

3. If you need ½ gallon or less to top up a large carboy that is still fermenting, prepare the following mixture:

½ gallon warm water (100°F [38°C])
3 cups sugar
2 tsp Vinacid O or R

Stir the mixture well to dissolve all the sugar, and add it to your carboy slowly. The mixture will ferment and maintain your acid and alcohol levels. Do not add it to a fined or filtered wine.

RECORD SHEET

Keep a detailed record sheet of your fresh grape wines.

FRESH GRAPES — FIRST RUN	EVALUATION AND NOTES
Date ______	______
Grape variety ______	______
Source ______	______
Amount ______	______
Grape cost ______	______
Starting SG ______	______
Starting acid ______	______
Adjustment SG ______	______
Adjustment acid ______	______
Yeast strain ______	______
1st racking date ______	______
2d racking date ______	______
3d racking date ______	______
Fining date ______	______
Chillproof date ______	______
Filter date ______	______
Bottling date ______	______
No. of bottles ______	______
Code for bottles ______	______

Vinifera Red — First Run

READY: 2 YEARS +

PRIMARY INGREDIENTS

12 lugs (420 lb)	Fresh vinifera red grapes	190.9 kg
4 pkts	Wine yeast (We suggest Montpellier.)	
4 tsp	Sulphite crystals	

SECONDARY INGREDIENTS

	Claro K. C. finings	
2 oz	Sulphite crystals	60 g

EQUIPMENT

Basic 10 + 45-gallon fermentor, 5 x 5-gallon carboys and 5 x 1-gallon jugs with bungs and fermentation locks to fit, 8-foot syphon hose, acid-testing kit, grape crusher, grape press, and destemmer; medium or coarse filter pads.

PRIMARY SEQUENCE

1. Crush the grapes into the primary fermentor. (See "Crushing and Pressing Grapes," page 249.)

2. Dissolve the sulphite crystals into 4 cups of warm water. Add the solution to the fermentor. Stir thoroughly with a paddle or a long-handled wooden or plastic spoon. Avoid inhaling the fumes.

3. Remove the grape stems with the destemmer paddle. Don't worry about every last stem.

4. Check the specific gravity (SG) of the must. It should be 1.090. If it is higher, reduce it by adding water. If it is lower, record it; you will need to add sugar after the first racking.

5. Check, and if necessary adjust, the acid content of the must. It should be 5.5–6.5 g/lit.

6. Check, and if necessary adjust, the temperature of the must. It should be 70–75°F (21–23°C).

7. Add the yeast to 4 cups of warm water. Let stand for 10 minutes. Stir in.

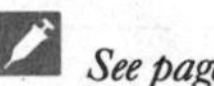

See page 220 *See page 223* *See page 224* *See page 226* *See page 227* *See page 229*

8. Cover the fermentor with a plastic sheet; tie down. After 24 hours, check that fermentation has begun. Foam should be visible on the surface and/or bubbles should be audible. If the fermentation has not begun, see "Stuck Ferment," page 269.

9. Stir twice daily to keep the floating cap wet.

10. Check SG daily. It should show an average daily drop of .010. You will be racking when it reaches SG 1.040. When it reaches 1.050, therefore, stop stirring — you will be racking the following day.

SECONDARY SEQUENCE

11. When SG falls to 1.040, rack into clean carboys or barrels. (See "Barrels," page 244.) Because the wine will be fermenting vigorously, fill the carboys only ¾ full (or the wine will foam up through the fermentation locks).

12. If your starting SG was too low and you need to add sugar, do so now. The addition of 3½ ounces (100 g) to 1 gallon (3.78 lit) will increase SG by .010. Dissolve the sugar in a little wine and pour it into the carboys or barrels.

13. Attach a fermentation lock to each carboy.

14. If you wish to make a second run, leave the pulp in the primary fermentor. If you do not wish to make a second run, press the pulp and syphon the liquid into gallon jugs. (See "Crushing and Pressing Grapes," page 249.) Use this liquid to top up after rackings.

15. After 10 days in secondary fermentors, rack. Top up.

16. If you are using barrels, check the aroma every week. When you smell the aroma of oak in the wine, syphon it into carboys.

17. After 6 months, add the finings. Let rest 10 days.

18. Rack into clean carboys. Add ¼ teaspoon of sulphite crystals to each 5 gallons, dissolved in a small amount of water. Top up.

19. If possible, chillproof to 32°F (0°C) for 2 weeks. (See "Chillproofing," page 248.)

20. Bulk age 9 months.

21. Filter.

22. Bottle.

23. Bottle age 1–5 years.

See page 232 (col. 1)  *See page 232 (col. 2)*  *See page 238* *See page 240 (col. 1)*  *See page 240 (col. 2)*

Vinifera Red — Second Run

READY: 3–12 MONTHS

PRIMARY INGREDIENTS

	Pulp from first run	
21.5 gal	Warm water (100–110°F [37–43°C])	81.4 lit
6 qt	Red grape concentrate	6 lit
32 lb	Sugar	14.5 kg
10 oz	Vinacid O or Acid Blend	300 g
10 tsp	Yeast nutrient	

SECONDARY INGREDIENTS

	Claro K. C. finings	
2 oz	Sulphite crystals	60 g

EQUIPMENT

Basic 10 + 45-gallon fermentor, 5 x 5-gallon carboys and 5 x 1-gallon jugs with bungs and fermentation locks to fit, 8-foot syphon hose, acid-testing kit, and grape press; medium or coarse filter pads

PRIMARY SEQUENCE

Begin the second run immediately after racking off your first run from the primary fermentor. At the bottom of the fermentor will be the pulp — grapeskins, seeds, a few stems, residual sugar, and a small yeast deposit.

1. Place all the primary ingredients in the primary fermentor. Stir thoroughly with a paddle or a long-handled wooden or plastic spoon until all the sugar is dissolved.

2. Do not adjust the starting specific gravity (SG). Because of the alcohol remaining from the first run, an accurate hydrometer reading will be impossible.

3. Check, and if necessary adjust, the temperature of the must. It should be 70–75°F (20–23°C).

4. Cover the fermentor with a plastic sheet; tie down. After 24 hours, check that fermentation has begun. Foam should be visible on the surface and/or bubbles should be audible. If the fermentation has not begun, see "Stuck Ferment," page 269.

5. Stir twice daily to keep the floating cap wet.

 *See page 220*  *See page 223* 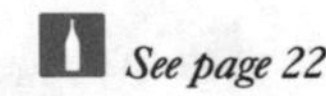 *See page 224* 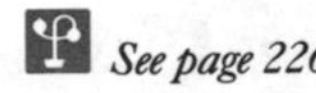 *See page 226* *See page 227* 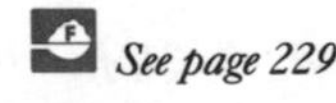 *See page 229*

6. Check SG daily. The fermentation will proceed much faster than in the first run. You will be racking when SG reaches 1.010. When it reaches 1.020, therefore, stop stirring — you will be racking the following day, if not sooner.

SECONDARY SEQUENCE

7. When SG falls to 1.010, rack into clean carboys.

8. Attach a fermentation lock to each carboy.

9. If you wish to make a third run, leave the pulp in the primary fermentor. If you do not wish to make a third run, press the pulp and syphon the liquid into gallon jugs. (See "Crushing and Pressing Grapes," page 249.) Use this liquid to top up after rackings.

10. After 10 days in carboys, rack into clean carboys. Top up.

11. After 6 weeks — or sooner if a firm deposit develops in the bottom of the carboys — rack into clean carboys.

12. Add the finings. Let rest 10 days.

13. Rack into clean carboys. Add ¼ teaspoon of sulphite crystals to each 5-gallon carboy, dissolved in a small amount of water. Top up.

14. If possible, chillproof to 32°F (0°C) for 2 weeks. (See "Chillproofing," page 248.)

15. Filter.

16. Bottle.

17. Bottle age 3 months.

If you wish to proceed with a third run, follow the same steps.

See page 232 (col. 1)  *See page 232 (col. 2)*  *See page 238*   *See page 240 (col. 1)*  *See page 240 (col. 2)*

Vinifera White — First Run

READY: 1 YEAR

PRIMARY INGREDIENTS

12 lugs (420 lb)	Fresh vinifera white grapes	190.9 kg
4 pkts	Champagne wine yeast	
4 tsp	Sulphite crystals	
6 tsp	Liquid tannin	

SECONDARY INGREDIENTS

	Claro K. C. or bentonite finings	
2 oz	Sulphite crystals	60 g

EQUIPMENT

Basic 10 + 45-gallon fermentor, 5 x 5-gallon carboys and 5 x 1-gallon jugs with bungs and fermentation locks to fit, 8-foot syphon hose, acid-testing kit, grape crusher, grape press, and destemmer; fine filter pads

PRIMARY SEQUENCE

1. Crush the grapes into the primary fermentor. (See "Crushing and Pressing Grapes," page 249.)

2. Dissolve the sulphite crystals in 4 cups of warm water. Add the solution to the fermentor. Stir thoroughly with a paddle or a long-handled wooden or plastic spoon. Avoid inhaling the fumes.

3. Remove the grape stems with the destemmer paddle. Don't worry about every last stem.

4. Check the specific gravity (SG) of the must. It should be 1.090. If it is higher, reduce it by adding water. If it is lower, record it; you will need to add sugar after the first racking.

5. Check, and if necessary adjust, the acid content of the must; it should be 6.5–7.5 g/lit.

6. Check, and if necessary adjust, the temperature of the must. It should be 70–75°F (21–23°C).

7. Add the yeast to 4 cups of warm water. Let stand for 10 minutes. Stir in.

See page 220  *See page 223* 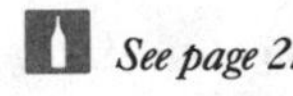 *See page 224*  *See page 226* 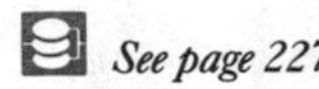 *See page 227* *See page 229*

8. Cover the fermentor with a plastic sheet; tie down. After 24 hours, whether fermentation has begun or not, remove the crushed grapes and stems and press them. (See "Crushing and Pressing Grapes," page 249.) Return the juice extracted by the pressing to the primary fermentor. Do not stir. If you wish to make a second run, place the pulp in a second primary fermentor. If you do not wish to make a second run, discard it.

9. After 48 hours, check that fermentation has begun. Foam should be visible on the surface and/or bubbles should be audible. If the fermentation has not begun, see "Stuck Ferment," page 269.

10. After 3 days, rack into the secondary fermentors. Because the wine will be vigorously fermenting, fill the carboys only ¾ full (or the wine will foam up through the fermentation locks).

11. Attach a fermentation lock to each carboy.

12. Check SG daily.

SECONDARY SEQUENCE

13. When SG falls to 1.000 or after 10 days, whichever comes first, rack into clean carboys. Top up.

14. If your starting SG was too low and you need to add sugar, do so now. The addition of 3½ ounces (100 g) to 1 gallon (3.78 lit) will increase the SG by .010. Dissolve the sugar in a little wine and pour it into the carboys.

15. After 6 weeks — or sooner if a firm deposit develops on the bottom of the carboys — rack into clean carboys. Add ¼ teaspoon of sulphite crystals to each 5-gallon carboy, dissolved in a small amount of water. Top up.

16. After 4 months, add the finings. Let rest 10 days.

17. Rack into clean carboys. Add ¼ teaspoon of sulphite crystals to each 5-gallon carboy, dissolved in a small amount of water. Top up.

18. If possible, chillproof to 32°F (0°C) for 2 weeks. (See "Chillproofing," page 248.)

19. Bulk age 6 months.

20. Filter.

21. Bottle.

22. Bottle age 1–2 years.

See page 232 (col. 1) 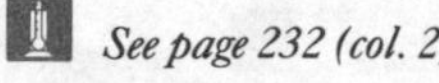 *See page 232 (col. 2)*  *See page 238*  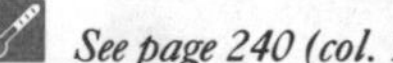 *See page 240 (col. 1)*  *See page 240 (col. 2)*

Vinifera White — Second Run

READY: 4 MONTHS

PRIMARY INGREDIENTS

	Pulp from first run	
21.5 gal	Warm water (100–110°F [37–43°C])	81.4 lit
6 qt	White grape concentrate	6 lit
4 pkts	Champagne wine yeast	
32 lb	Sugar	14.5 kg
1¾ cups	Vinacid O or Acid Blend	420 g
5 tsp	Yeast nutrient	

SECONDARY INGREDIENTS

	Claro K. C. or bentonite finings	
2 oz	Sulphite crystals	60 g

EQUIPMENT

Basic 10 + 45-gallon fermentor, 5 x 5-gallon carboys and 5 x 1-gallon jugs with bungs and fermentation locks to fit, 8-foot syphon hose, acid-testing kit, and grape press; fine filter pads

PRIMARY SEQUENCE

Begin the second run immediately after pressing your first run.

1. Place the pulp from the first run, together with the other primary ingredients, in the primary fermentor. Stir thoroughly until all the sugar is dissolved.

2. Check, and if necessary adjust, the specific gravity (SG) of the must. It should be 1.085.

3. Check, and if necessary adjust, the temperature of the must. It should be 70–75°F (21–23°C).

4. Cover the fermentor with a plastic sheet; tie down. After 24 hours, check that fermentation has begun. Foam should be visible on the surface and/or bubbles should be audible. If the fermentation has not begun, see "Stuck Ferment," page 269.

5. Stir twice daily to keep the floating cap wet.

6. Check SG daily. The fermentation will proceed much faster than in the first run. You will be racking when SG reaches 1.040. When it reaches 1.050, therefore, stop

 *See page 220*  *See page 223*  *See page 224*   *See page 226*  *See page 227*  *See page 229*

stirring — you will be racking the following day, if not sooner.

SECONDARY SEQUENCE

7. When SG falls to 1.040, rack into clean carboys. Because the wine will be fermenting vigorously, fill the carboys only ¾ full (or the wine will foam up through the fermentation locks).

8. Attach a fermentation lock to each carboy.

9. If you wish to make a third run, leave the pulp in the primary fermentor. If you do not wish to make a third run, press the pulp and syphon the liquid into gallon jugs. (See "Crushing and Pressing Grapes," page 249.) Use this liquid to top up after rackings.

10. After 10 days in carboys, rack into clean carboys. Top up.

11. After 6 weeks — or sooner if a firm deposit develops in the bottom of the carboys — rack into clean carboys.

12. Add the finings. Let rest 10 days.

13. Rack into clean carboys. Add ¼ teaspoon of sulphite crystals to each 5-gallon carboy, dissolved in a small amount of water. Top up.

14. If possible, chillproof to 32°F (0°C) for 2 weeks. (See "Chillproofing," page 248.)

15. Filter.

16. Bottle.

17. Bottle age 3 months.

If you wish to proceed with a third run, follow the same steps.

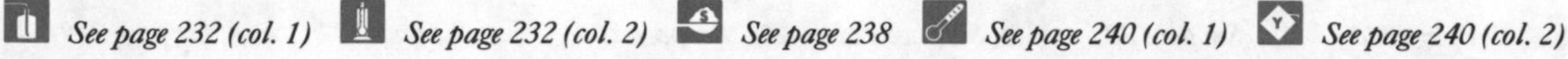
See page 232 (col. 1) *See page 232 (col. 2)*
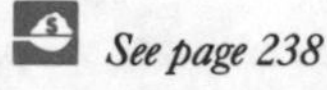
See page 238
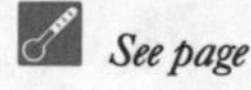
See page 240 (col. 1)

See page 240 (col. 2)

Hybrid Red Ameliorated with Grape Concentrate

READY: 2–3 YEARS

PRIMARY INGREDIENTS

200 lb	Fresh hybrid red grapes	91 kg
2 tsp	Sulphite crystals	
2 gal	Red grape concentrate	8 lit
4 gal	HOT water	16 lit
13 lb	Sugar	5.9 kg
4 gal	COLD water	16 lit
10 tsp	Pectic enzyme	
5 tsp	Yeast nutrient	
4 pkts	Montpellier wine yeast	

SECONDARY INGREDIENTS

	Bentonite finings	
2 oz	Sulphite crystals	60 g

EQUIPMENT

Basic 10 + 45-gallon fermentor, 5 x 5-gallon carboys and 5 x 1-gallon jugs with bungs and fermentation locks to fit, 8-foot syphon hose, acid-testing kit, grape crusher, grape press, and destemmer; coarse filter pads

Note: A second run is not recommended.

PRIMARY SEQUENCE

1. Crush the grapes into the primary fermentor. (See "Crushing and Pressing Grapes," page 249.)

2. Remove the grape stems with the destemmer paddle. Don't worry about every last stem.

3. Add the grape concentrate.

4. Dissolve the sulphite crystals in 2 cups of warm water. Add the solution to the fermentor. Stir thoroughly with a paddle or a long-handled wooden or plastic spoon. Avoid inhaling the fumes.

5. Dissolve the sugar in 4 gallons of hot tap water and add it to the primary fermentor. Stir well.

6. Add 4 gallons of cold tap water. Stir well.

7. Add the yeast nutrient and the pectic enzyme. Stir well.

8. Check the specific gravity (SG) of the must. It should be 1.090. If it is higher, reduce it by adding water. If it is lower,

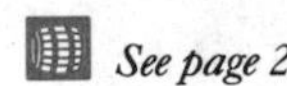

See page 220 See page 223 See page 224 See page 226 See page 227 See page 229

record it; you will need to add sugar after the first racking.

9. Check, and if necessary adjust, the acid content of the must; it should be 5.5–6.5 g/lit.

10. Check, and if necessary adjust, the temperature of the must. It should be 70–75°F (21–23°C).

11. After 4 hours, add the yeast to 4 cups of warm water. Let stand for 10 minutes. Stir in.

12. Cover the fermentor with a plastic sheet; tie down. After 24 hours, check that fermentation has begun. Foam should be visible on the surface and/or bubbles should be audible. If the fermentation has not begun, see "Stuck Ferment," page 269.

13. Stir twice daily to keep the floating cap wet.

14. Check SG daily. It should show an average daily drop of .010. You will be racking when it reaches SG 1.020. When it reaches 1.030, therefore, stop stirring — you will be racking the following day.

SECONDARY SEQUENCE

15. When SG falls to 1.020, rack into clean carboys. Because the wine will be fermenting vigorously, fill the carboys only ¾ full (or the wine will foam up through the fermentation locks).

16. If your starting SG was too low and you need to add sugar, do so now. The addition of 3½ ounces (100 g) to 1 gallon (3.78 lit) will increase SG by .010. Dissolve the sugar in a little wine and pour it into the carboys.

17. Attach a fermentation lock to each carboy.

18. Press the pulp and syphon the liquid into gallon jugs. (See "Crushing and Pressing Grapes," page 249.) Use this liquid to top up after rackings.

19. After 10 days in carboy, rack into clean carboys. Rack again in 3 weeks. Add ¼ teaspoon of sulphite crystals to each 5-gallon carboy, dissolved in a small amount of water. Top up.

20. After 6 months, add the finings. Let rest 10 days.

21. Rack into clean carboys. Add ¼ teaspoon sulphite crystals to each 5-gallon carboy, dissolved in a small amount of water. Top up.

22. If possible, chillproof to 32°F (0°C) for 2 weeks. (See "Chillproofing," page 248.)

23. Bulk age 9 months.

24. Filter.

25. Bottle.

26. Bottle age 1–5 years.

See page 232 (col. 1) *See page 232 (col. 2)* *See page 238* *See page 240 (col. 1)* *See page 240 (col. 2)*

Hybrid Red Ameliorated with Sugar and Water

READY: 2 YEARS

This "economy" recipe for red hybrids makes a thin red wine.

PRIMARY INGREDIENTS

300 lb	Fresh hybrid red grapes	136.4 kg
3 tsp	Sulphite crystals	
5 gal	HOT water	20 lit
25 lb	Sugar	11.4 kg
4 gal	COLD water	16 lit
10 tsp	Pectic enzyme	
10 tsp	Yeast nutrient	
4 pkts	Montpellier wine yeast	

SECONDARY INGREDIENTS

	Bentonite finings	
2 oz	Sulphite crystals	60 g

EQUIPMENT

Basic 10 + 45-gallon fermentor, 5 x 5-gallon carboys and 5 x 1-gallon jugs with bungs and fermentation locks to fit, 8-foot syphon hose, acid-testing kit, grape crusher, grape press, and destemmer; coarse filter pads

Note: A second run is not recommended.

PRIMARY SEQUENCE

1. Crush the grapes into the primary fermentor. (See "Crushing and Pressing Grapes," page 249.)

2. Remove the grape stems with the destemmer paddle. Don't worry about every last one.

3. Dissolve the sulphite crystals in 3 cups of warm water. Add the solution to the fermentor. Stir thoroughly with a paddle or a long-handled wooden or plastic spoon. Avoid inhaling the fumes.

4. Dissolve the sugar in 5 gallons of hot tap water and add it to the primary fermentor. Stir well.

5. Add 4 gallons of cold water. Stir well.

6. Add the yeast nutrient and the pectic enzyme. Stir well.

7. Check the specific gravity (SG) of the must. It should be 1.090. If it is higher, lower it by adding water. If it is lower, record it; you will need to add sugar after the first racking.

See page 220  See page 223  See page 224  See page 226  See page 227  See page 229

8. Check, and if necessary adjust, the acid content of the must; it should be 7.5 g/lit.

9. Check, and if necessary adjust, the temperature of the must. It should be 70–75°F (21–23°C).

10. After 4 hours, add the yeast to 4 cups of warm water. Let stand for 10 minutes. Stir in.

11. Cover the fermentor with a plastic sheet; tie down. After 24 hours, check that fermentation has begun. Foam should be visible on the surface and/or bubbles should be audible. If the fermentation has not begun, see "Stuck Ferment," page 269.

12. Stir twice daily to keep the floating cap wet.

13. Check SG daily. It should show an average daily drop of .010. You will be racking when it reaches SG 1.020. When it reaches 1.030, therefore, stop stirring — you will be racking the following day.

SECONDARY SEQUENCE

14. When SG falls to 1.020, rack into clean carboys. Because the wine will be fermenting vigorously, fill the carboys only ¾ full (or the wine will foam up through the fermentation locks).

15. If your starting SG was too low and you need to add sugar, do so now. The addition of 3½ ounces (100 g) to 1 gallon (3.78 lit) will increase SG by .010. Dissolve the sugar in a little wine and pour it into the carboys.

16. Attach a fermentation lock to each carboy.

17. Press the pulp and syphon the liquid into gallon jugs. (See "Crushing and Pressing Grapes," page 249.) Use this liquid to top up after rackings.

18. After 10 days in carboys, rack into clean carboys. Rack again in 3 weeks. Add ¼ teaspoon of sulphite crystals to each 5-gallon carboy, dissolved in a small amount of water. Top up.

19. After 6 months, add the finings. Let rest 10 days.

20. Rack into clean carboys. Add ¼ teaspoon of sulphite crystals to each 5-gallon carboy, dissolved in a small amount of water. Top up.

21. If possible, chillproof to 32°F (0°C) for 2 weeks. (See "Chillproofing," page 248.)

22. Bulk age 9 months.

23. Filter.

24. Bottle.

25. Bottle age 1–5 years.

See page 232 (col. 1)  *See page 232 (col. 2)*  *See page 238*  *See page 240 (col. 1)*  *See page 240 (col. 2)*

Hybrid Red Ameliorated with Sugar and Chalk

READY: 1½–3 YEARS

Deacidifying juice with an alkali is not our favorite procedure, and we have yet to taste a truly excellent wine made by this means — but it is cheaper than adding grape concentrate.

PRIMARY INGREDIENTS

400 lb	Fresh hybrid red grapes	181.8 kg
3 tsp	Sulphite crystals	
1 lb	Chalk	450 g
11 lb	Sugar, or quantity as required	5 kg
10 tsp	Pectic enzyme	
5 tsp	Yeast nutrient	
4 pkts	Montpellier wine yeast	

SECONDARY INGREDIENTS

	Bentonite finings	
2 oz	Sulphite crystals	60 g

EQUIPMENT

Basic 10 + 45-gallon fermentor, 5 x 5-gallon *(continues)* carboys and 5 x 1-gallon jugs with bungs and fermentation locks to fit, 8-foot syphon hose, acid-testing kit, grape crusher, grape press, and destemmer; coarse filter pads

Note: A second run is not recommended.

PRIMARY SEQUENCE

1. Crush the grapes into the primary fermentor. (See "Crushing and Pressing Grapes," page 249.)

2. Remove the grape stems with the destemmer paddle. Don't worry about every last one.

3. Dissolve the sulphite crystals in 3 cups of warm water. Add the solution to the fermentor. Stir thoroughly with a paddle or a long-handled wooden or plastic spoon. Avoid inhaling the fumes.

4. Strain off 5 gallons (20 lit) of juice into your regular (8-gallon) primary fermentor. Stir in the chalk. Stir every 2–4 hours for 12 hours, then syphon back into the main fermentor, taking care to leave the sediment behind.

5. Add the yeast nutrient and pectic enzyme. Stir well.

 *See page 220*  *See page 223*  *See page 224*  *See page 226*  *See page 227*  *See page 229*

6. Check the specific gravity (SG) of the must. It should be 1.090. If it is higher, reduce it by adding water. If it is lower, record it; you will need to add sugar after the first racking.

7. Check, and if necessary adjust, the acid content of the must; it should be 7.5 g/lit, maximum.

8. Check, and if necessary adjust, the temperature of the must. It should be 70–75°F (21–23°C).

9. After 4 hours, add the yeast to 4 cups of warm water. Let stand for 10 minutes. Stir in.

10. Cover the fermentor with a plastic sheet; tie down. After 24 hours, check that fermentation has begun. Foam should be visible on the surface and/or bubbles should be audible. If the fermentation has not begun, see "Stuck Ferment," page 269.

11. Stir twice daily to keep the floating cap wet.

12. Check SG daily. It should show an average daily drop of .010. You will be racking when it reaches SG 1.040. When it reaches 1.050, therefore, stop stirring — you will be racking the following day. Note: If you want a deeper color to the wine, you may delay racking until SG 1.030 or 1.020.

SECONDARY SEQUENCE

13. When SG falls to 1.040, rack into clean carboys. Because the wine will be fermenting vigorously, fill the carboys only ¾ full (or the wine will foam up through the fermentation locks).

14. If your starting SG was too low and you need to add sugar, do so now. The addition of 3½ ounces (100 g) to 1 gallon (3.78 lit) will increase SG by .010. Dissolve the sugar in a quantity of wine and pour it into the carboys.

15. Attach a fermentation lock to each carboy.

16. Press the pulp and syphon the liquid into gallon jugs. (See "Crushing and Pressing Grapes," page 249.) Use this liquid to top up after rackings.

17. After 10 days in carboys, rack into clean carboys. Top up. Rack again in 3 weeks. Add ¼ teaspoon of sulphite crystals to each 5-gallon carboy, dissolved in a small amount of water. Top up.

18. After 6 months, add the finings. Let rest 10 days.

19. Rack into clean carboys. Add ¼ teaspoon of sulphite crystals to each 5-gallon carboy, dissolved in a small amount of water. Top up.

20. If possible, chillproof to 32°F (0°C) for 2 weeks. (See "Chillproofing," page 248.)

21. Bulk age 3 months.

22. Filter.

23. Bottle.

24. Bottle age 1–3 years.

See page 232 (col. 1)  *See page 232 (col. 2)*  *See page 238* *See page 240 (col. 1)*  *See page 240 (col. 2)*

Hybrid White Ameliorated with Grape Concentrate — First Run

READY: 1½–2½ YEARS

The high acid and low body of hybrids means they make a better white wine than red. While we do not recommend a second run on reds, a 10-15 gallon second run on white hybrids is worthwhile.

PRIMARY INGREDIENTS

200 lb	Fresh hybrid white grapes	90.1 kg
6 qt	White grape concentrate	6 lit
4 gal	HOT water	16 lit
2 tsp	Sulphite crystals	
4 gal	COLD tap water	16 lit
14 lb	Sugar	6.4 kg
10 tsp	Pectic enzyme	
10 tsp	Yeast nutrient	
5 tsp	Liquid tannin	
4 pkts	Champagne wine yeast	

SECONDARY INGREDIENTS

	Claro K. C. finings	
2 oz	Sulphite crystals	60 g
1½ qt	Wine conditioner	1.5 lit

EQUIPMENT

Basic 10 + 45-gallon fermentor, 5 x 5-gallon carboys and 5 x 1-gallon jugs with bungs and fermentation locks to fit, 8-foot syphon hose, acid-testing kit, grape crusher, grape press, destemmer; fine filter pads

PRIMARY SEQUENCE

1. Crush the grapes into the primary fermentor. (See "Crushing and Pressing Grapes," page 249.)

2. Remove the grape stems with the destemmer paddle. Don't worry about every last one.

3. Dissolve the sulphite crystals in 3 cups of warm water. Add the solution to the fermentor. Stir thoroughly with a paddle or a long-handled wooden or plastic spoon. Avoid inhaling the fumes.

4. Dissolve the sugar in 4 gallons of hot tap water. Add it to the primary fermentor. Stir well.

5. Add 4 gallons of cold tap water. Stir well.

6. Add the yeast nutrient, the pectic enzyme, and the liquid tannin. Stir well.

7. Check the specific gravity (SG) of the must. It should be 1.090. If it is

 See page 220  See page 223  See page 224  See page 226  See page 227 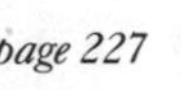 See page 229

higher, reduce it by adding water. If it is lower, record it; you will need to add sugar after the first racking.

8. Check, and if necessary adjust, the acid content of the must; it should be 7 g/lit. However, if you have 8 g/lit, do not adjust; chillproofing and sweetening will take care of it.

9. Check, and if necessary adjust, the temperature of the must. It should be 70–75°F (21–23°C).

10. After 4 hours, add the yeast to 4 cups of warm water. Let stand for 10 minutes. Stir in.

11. Cover the fermentor with a plastic sheet; tie down. After 24 hours, whether fermentation has begun or not, remove the crushed grapes and stems and press them. (See “Crushing and Pressing Grapes,” page 249.) Return the juice extracted by the pressing to the primary fermentor. Do not stir. If you wish to make a second run, place the pulp in a second primary fermentor. If you do not wish to make a second run, discard it.

12. After 48 hours, check that fermentation has begun. Foam should be visible on the surface and/or bubbles should be audible. If the fermentation has not begun, see “Stuck Ferment,” page 269.

13. After 4–5 days, rack into clean carboys. Because the wine will be vigorously fermenting, fill the carboys only ¾ full (or the wine will foam up through the fermentation locks).

14. Attach a fermentation lock to each carboy.

15. Check SG daily.

SECONDARY SEQUENCE

16. When SG falls to 1.020, rack into clean carboys. Top up.

17. If your starting SG was too low and you need to add sugar, do so now. The addition of 3½ ounces (100 g) to 1 gallon (3.78 lit) will increase SG by .010. Dissolve the sugar in a little wine and pour it into the carboys. After 10 days, rack into clean carboys. Top up. Rack again in 3 weeks. Add ¼ teaspoon sulphite crystals to each 5-gallon carboy, dissolved in a small amount of water. Top up.

18. After 6 weeks — or sooner if a firm deposit develops on the bottom of the carboys — rack into clean carboys. Add ¼ teaspoon of sulphite to each 5-gallon carboy, dissolved in a small amount of water. Top up.

19. After 4 months, add the finings. Let rest 10 days.

20. Rack into clean carboys. Add ¼ teaspoon sulphite to each 5-gallon carboy, dissolved in a small amount of water. Top up.

21. If possible, chillproof to 32°F (0°C) for 2 weeks. (See “Chillproofing,” page 248.)

22. Bulk age 6 months.

23. Filter.

24. Sweeten to taste with wine conditioner and bottle.

25. Bottle age 1–2 years.

See page 232 (col. 1)  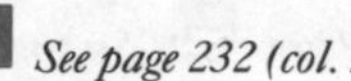 *See page 232 (col. 2)*  *See page 238*  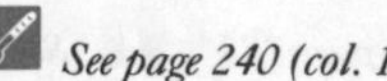 *See page 240 (col. 1)*  *See page 240 (col. 2)*

Hybrid White Ameliorated with Grape Concentrate — Second Run

READY: 9 MONTHS

PRIMARY INGREDIENTS

	Pulp from first run	
5 gal	HOT water	20 lit
3 qt	White grape concentrate	3 lit
20 lb	Sugar	9 kg
5 gal	COLD water	20 lit
5 tsp	Liquid tannin	
5 tsp	Yeast nutrient	

SECONDARY INGREDIENTS

	Claro K. C. finings	
2 oz	Sulphite crystals	60 g
1½ qt	Wine conditioner	1.5 lit

EQUIPMENT

Basic 10 + 45-gallon fermentor, 5 x 5-gallon carboys and 5 x 1-gallon jugs with bungs and fermentation locks to fit, 8-foot syphon hose, acid-testing kit, and grape press; fine filter pads

PRIMARY SEQUENCE

Begin the second run immediately after pressing your first run.

1. Dissolve the sugar in 5 gallons of hot tap water. Add it to the primary fermentor.

2. Add the grape concentrate and stir well with a long-handled wooden or plastic spoon.

3. Add 5 gallons of cold tap water.

4. Add the liquid tannin and the yeast nutrient. Stir well.

5. Do not adjust the starting specific gravity (SG). Because of the alcohol remaining from the first run, an accurate hydrometer reading will be impossible.

6. Cover the fermentor with a plastic sheet; tie down. After 24 hours, check that fermentation has begun. Foam should be visible on the surface and/or bubbles should be audible. If the fermentation has not begun, see "Stuck Ferment," page 269.

7. Stir twice daily to keep the floating cap wet.

8. Check SG daily. The fermentation will proceed much faster than in the first run. You will be racking when SG reaches

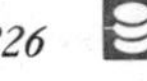

See page 220 See page 223 See page 224 See page 226 See page 227 See page 229

1.010. When it reaches 1.020, therefore, stop stirring — you will be racking the following day, if not sooner.

SECONDARY SEQUENCE

9. When SG falls to 1.010, rack into clean carboys. Top up.

10. Attach a fermentation lock to each carboy.

11. Press the pulp lightly and syphon the liquid into gallon jugs. (See "Crushing and Pressing Grapes," page 249.) Use this liquid to top up after rackings.

12. After 10 days in carboys, rack into clean carboys. Top up. Rack again in 3 weeks. Add ¼ teaspoon of sulphite crystals to each 5-gallon carboy, dissolved in a small amount of water. Top up.

13. After 6 weeks — or sooner if a firm deposit develops in the bottom of the carboys — rack into clean carboys.

14. Add the finings. Let rest 10 days.

15. Rack into clean carboys. Add ¼ teaspoon of sulphite crystals to each 5-gallon carboy, dissolved in a small amount of water. Top up.

16. If possible, chillproof to 32°F (0°C) for 2 weeks. (See "Chillproofing," page 248.)

17. Filter.

18. Sweeten to taste with wine conditioner and bottle.

19. Bottle age 6 months.

We recommend that you blend your second-run wine with your first run to achieve a more balanced wine.

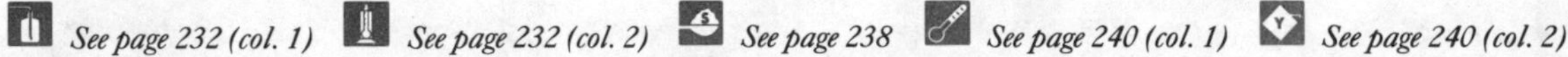
See page 232 (col. 1)  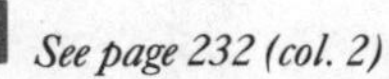 *See page 232 (col. 2)*  *See page 238*  *See page 240 (col. 1)*  *See page 240 (col. 2)*

Hybrid White Ameliorated with Sugar and Water — First Run

READY: 6 MONTHS

Although it is difficult to achieve a good red wine with hybrid grapes ameliorated with sugar and water, this method of amelioration works well with white wines, which require less body.

PRIMARY INGREDIENTS

300 lb	Fresh hybrid white grapes	136.4 kg
3 tsp	Sulphite crystals	
5 gal	HOT water	20 lit
10 lb	Sugar	4.5 kg
10 tsp	Pectic enzyme	
10 tsp	Yeast nutrient	
5 tsp	Liquid tannin	
4 pkts	Champagne wine yeast	

SECONDARY INGREDIENTS

	Claro K. C. finings	
2 tsp	Sulphite crystals	
1½ qt	Wine conditioner	1.5 lit
1 oz	Sinatin 17	30 ml

EQUIPMENT

Basic 10 + 45-gallon fermentor, 5 x 5-gallon carboys and 5 x 1-gallon jugs with bungs and fermentation locks to fit, 8-foot syphon hose, acid-testing kit, grape crusher, grape press, and destemmer; fine filter pads

PRIMARY SEQUENCE

1. Crush the grapes into the primary fermentor. (See "Crushing and Pressing Grapes," page 249.)

2. Remove the grape stems with the destemmer paddle. Don't worry about every last one.

3. Dissolve the sulphite crystals in 3 cups of warm water. Add the solution to the fermentor. Stir thoroughly with a paddle or a long-handled wooden or plastic spoon. Avoid inhaling the fumes.

4. Dissolve the sugar in 5 gallons of hot tap water and add it to the primary fermentor. Stir well.

5. Add the yeast nutrient, the pectic enzyme, and the liquid tannin. Stir well.

6. Check the specific gravity (SG) of the must. It should be 1.090. If it is

 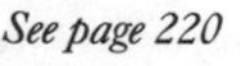 See page 220  See page 223  See page 224  See page 226  See page 227  See page 229

higher, reduce it by adding water. If it is lower, record it; you will need to add sugar after the first racking.

7. Check, and if necessary adjust, the acid content of the must; it should be 7.5 g/lit. However, if you have 8 g/lit, do not adjust; chillproofing and sweetening will take care of it.

8. Check, and if necessary adjust, the temperature of the must. It should be 70–75°F (21–23°C).

9. After 4 hours, add the yeast to 4 cups of warm water. Let stand for 10 minutes. Stir in.

10. Cover the fermentor with a plastic sheet; tie down. After 24 hours, whether fermentation has begun or not, remove the crushed grapes and stems and press them. Return the juice extracted by the pressing to the primary fermentor. Do not stir. If you wish to make a second run, place the pulp in a second primary fermentor. If you do not wish to make a second run, discard the pulp.

11. After 48 hours, check that fermentation has begun. Foam should be visible on the surface and/or bubbles should be audible. If the fermentation has not begun, see "Stuck Ferment," page 269.

12. After 3 days, rack into the secondary fermentors. Because the wine will be vigorously fermenting, fill the carboys only ¾ full (or the wine will foam up through the fermentation locks).

13. Attach a fermentation lock to each carboy.

14. Check SG daily.

SECONDARY SEQUENCE

15. When SG falls to 1.020, rack into clean carboys. Top up.

16. If your starting SG was too low and you need to add sugar, do so now. The addition of 3½ ounces (100 g) to 1 gallon (3.78 lit) will increase the SG by .010. Dissolve the sugar in a little wine and pour it into the carboys.

17. After 10 days, rack into clean carboys. Top up. Rack again in 3 weeks. Add ¼ teaspoon sulphite crystals to each 5-gallon carboy, dissolved in a small amount of water. Top up.

18. After 6 weeks — or sooner if a firm deposit develops on the bottom of the carboy — rack into clean carboys. Top up.

19. After 4 months, add the finings. Let rest 10 days.

20. Rack into clean carboys. Add ¼ teaspoon of sulphite crystals to each 5-gallon carboy, dissolved in a small amount of water. Top up.

21. If possible, chillproof to 32°F (0°C) for 2 weeks. (See "Chillproofing," page 248.)

22. Add the Sinatin 17 and bulk age 6 months.

23. Filter.

24. Add 2 ounces of wine conditioner per gallon and bottle.

25. Bottle age 1–2 years.

See page 232 (col. 1)  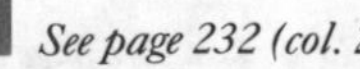 *See page 232 (col. 2)*  *See page 238*  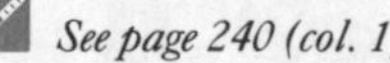 *See page 240 (col. 1)*  *See page 240 (col. 2)*

Hybrid White Ameliorated with Sugar and Water — Second Run

READY: 3–6 MONTHS

This second run may well be better than the first run.

PRIMARY INGREDIENTS

	Pulp from first run	
5 gal	HOT water	20 lit
3 qt	White grape concentrate	3 lit
20 lb	Sugar	9 kg
5 gal	COLD water	20 lit
5 tsp	Liquid tannin	
5 tsp	Yeast nutrient	

SECONDARY INGREDIENTS

	Claro K. C. or bentonite finings	
2 oz	Sulphite crystals	60 g
1 oz	Sinatin 17	30 ml
20 oz	Wine conditioner	600 ml

EQUIPMENT

Basic 10 + 45-gallon fermentor, 2 x 5-gallon carboys and 5 x 1-gallon jugs with bungs and fermentation locks to fit, 8-foot syphon hose, acid-testing kit, and grape press; fine filter pads

PRIMARY SEQUENCE

Begin the second run immediately after pressing your first run.

1. Dissolve the sugar in 5 gallons of hot tap water. Add it to the primary fermentor.

2. Add the grape concentrate and stir well with a long-handled wooden or plastic spoon.

3. Add 5 gallons of cold tap water.

4. Add the liquid tannin and the yeast nutrient. Stir well.

5. Do not adjust the starting specific gravity (SG). Because of the alcohol remaining from the first run, an accurate hydrometer reading will be impossible.

6. Cover the fermentor with a plastic sheet; tie down. After 24 hours, check that fermentation has begun. Foam should be visible on the surface and/or bubbles should be audible. If the fermentation has not begun, see "Stuck Ferment," page 269.

7. Stir twice daily to keep the floating cap wet.

8. Check SG daily. The fermentation will proceed much faster than in the first run. You will be racking when SG reaches

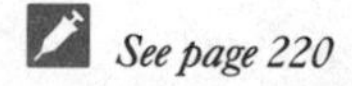

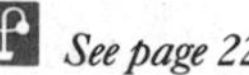

See page 220 See page 223 See page 224 See page 226 See page 227 See page 229

1.020. When it reaches 1.030, therefore, stop stirring — you will be racking the following day, if not sooner.

SECONDARY SEQUENCE

9. When SG falls to 1.020, rack into clean carboys. Because the wine will be fermenting vigorously, fill the carboys only ¾ full (or the wine will foam up through the fermentation locks).

10. Attach a fermentation lock to each carboy.

11. Press the pulp lightly and syphon the liquid into gallon jugs. (See "Crushing and Pressing Grapes," page 249.) Use this liquid to top up after rackings.

12. After 10 days, rack into clean carboys. Top up. Rack again in 3 weeks. Add ¼ teaspoon of sulphite crystals to each 5-gallon carboy, dissolved in a small amount of water. Top up.

13. After 6 weeks — or sooner if a firm deposit develops in the bottom of the carboys — rack into clean carboys.

14. Add the finings. Let rest 10 days.

15. Rack into clean carboys. Add ¼ teaspoon of sulphite crystals to each 5-gallon carboy, dissolved in a small amount of water. Top up.

16. If possible, chillproof to 32°F (0°C) for 2 weeks. (See "Chillproofing," page 248.)

17. Filter.

18. Add wine conditioner and the Sinatin. Bottle.

19. Bottle age 3 months.

We recommend that you blend your second-run wine with your first run to achieve a more balanced wine.

See page 232 (col. 1)  *See page 232 (col. 2)*  *See page 238*  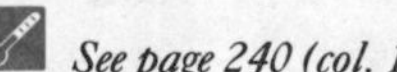 *See page 240 (col. 1)*  *See page 240 (col. 2)*

Wine from Table Grapes

READY: 7 MONTHS

Thompson seedless table grapes are frequently used to make lower-priced wines. If you have access to these grapes economically, they make a Chablis-style, soft wine.

PRIMARY INGREDIENTS

App 3 lugs (90 lb)	Thompson seedless grapes	41 kg
2 lb	Sugar	1 kg
5 tsp	Vinacid R	
2 tsp	Pectic enzyme	
2 tsp	Liquid tannin	
8	Campden tablets (crushed)	
1 pkt	Champagne wine yeast	

SECONDARY INGREDIENTS

	Bentonite finings	
¼ tsp	Sulphite crystals	
8 oz	Wine conditioner	240 ml

EQUIPMENT

Basic 10 + grape crusher and press; fine filter pads

PRIMARY SEQUENCE

1. Crush the grapes into the primary fermentor. (See "Crushing and Pressing Grapes," page 249.)

2. Remove the grape stems with the destemmer paddle.

3. Add the sugar and the Vinacid R. Stir thoroughly until all the sugar is dissolved.

4. Add the next 3 ingredients. Mix well.

5. Check, and if necessary adjust, the specific gravity (SG) of the must. It should be 1.095.

6. Check, and if necessary adjust, the temperature of the must. It should be 75°F (23°C).

7. Add the yeast to a cup of warm water. Let stand for 10 minutes. Stir in.

8. Cover the fermentor with a plastic sheet; tie down. Keep in a warm place (75°F [23°C]). After 24 hours, check that fermentation has begun. Foam should be visible on the surface and/or bubbles should be audible. If fermentation has not begun, see "Stuck Ferment," page 269.

9. After 24 hours of fermentation, remove the crushed grapes and stems and press them.

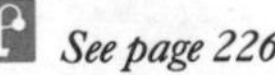

See page 220 See page 223 See page 224 See page 226 See page 227 See page 229

Return the juice extracted by the pressing to the primary fermentor. Discard the pulp.

10. Check SG every other day.

SECONDARY SEQUENCE

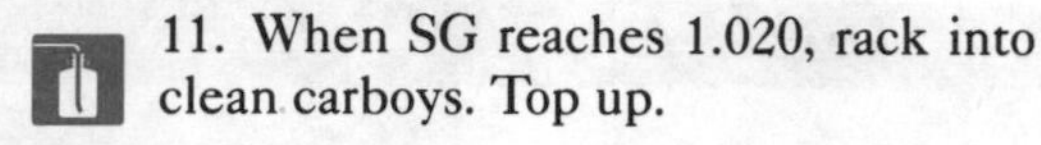

11. When SG reaches 1.020, rack into clean carboys. Top up.

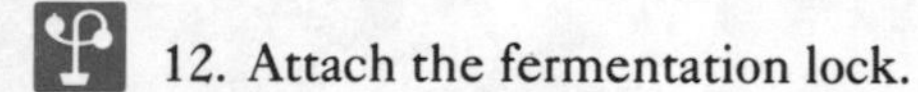

12. Attach the fermentation lock.

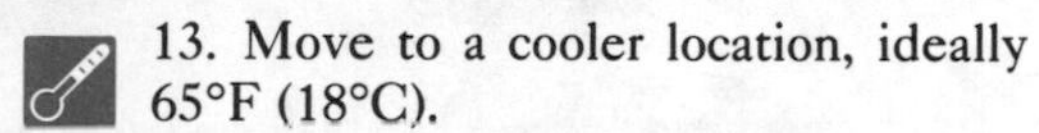

13. Move to a cooler location, ideally 65°F (18°C).

14. After 10 days or at SG 1.000, whichever comes first, rack into a clean carboy. Top up.

15. After 3 weeks or at SG .990–.995, whichever comes first, rack into a clean carboy.

16. Add the finings. Let rest 10 days.

17. Rack into the primary fermentor.

18. Filter into a clean carboy.

19. Add ¼ teaspoon of sulphite crystals dissolved in a small amount of tap water. Top up.

20. Bulk age 1 month.

21. Add the wine conditioner and bottle.

22. Bottle age 5 months.

See page 232 (col. 1)  *See page 232 (col. 2)*  *See page 238* 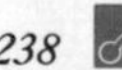 *See page 240 (col. 1)*  *See page 240 (col. 2)*

CHAMPAGNE

Champagne — Méthode Champenoise

The méthode champenoise *is the traditional method of making champagne, and it is the most complicated. The procedure requires a great deal of dedication, practice, and technique to pull off.*

STEP ONE — *TIRAGE*

Ingredients

5 gal (20 lit) *cuvée* — sound, dry white wine with an alcohol content of 10.5% and an acid content of 6.5–7.0 g/lit. It should be 6 months old, fined, filtered, and chillproofed.

13 oz (390 g) white cane or beet sugar
5 tsp yeast nutrient
2 x 5-g packets champagne yeast

Equipment

33 oz wine bottle with a plastic cap to fit
1 liter glass pitcher
5 gallon carboy
Long-handled wooden or plastic stirring spoon
Syphon hose

Procedure

1. Syphon off 33 oz (1 lit) of *cuvée* into the wine bottle. Seal it with a plastic cap and set it aside.

2. Syphon off 16 oz (480 ml) of *cuvée* into the glass pitcher. Add the sugar and the yeast nutrient; stir until both solids have dissolved.

3. Add the yeast to 1 cup of warm water (100°F [37°C]). Let it stand 10 minutes.

4. Pour the sugar and yeast nutrient solution into the carboy. Add the yeast.

5. Syphon the remaining *cuvée* into the carboy. Stir with the long handle of the spoon. Attach a fermentation lock.

STEP TWO

Equipment

25 x 25-oz (750-ml) champagne bottles
26 crown caps (good quality)
Crown capper

Procedure

6. After 24–48 hours, when the *cuvée* is cloudy with growing yeast, assemble the champagne bottles. Wash them, but do not sulphite them.

7. Remove the fermentation lock from the *cuvée* carboy and stir the yeast into suspension with the long handle of the spoon.

8. Syphon the *cuvée* into the champagne bottles, leaving 1 inch (2.5 cm) of air space at the top of each bottle. Crown-cap the bottles.

9. Store the bottles for 3 months in an upright position at 65–75°F (19–23°C). Twice a week during this time, invert each bottle, shake it gently to encourage fermentation, and return it to an upright position. A substantial yeast sediment will develop in the bottles.

STEP THREE — *REMUAGE*

Caution: The bottles are now under pressure and dangerous. From this point on, always wear safety gloves and eye protection when handling.

Equipment

Eye protection
Safety gloves
Riddling board (See diagram, page 210.)

Procedure

10. Place the bottles carefully in the riddling board. Every day, give each bottle a quarter turn with a jolting motion. This will shake the yeast sediment down onto the cap of the bottle. Continue this for 6 weeks. At the beginning, arrange the riddling board so that it is angled steeply. Week by week, open it out so that the bottles are tilted more and more toward the vertical.

STEP FOUR — *DÉGORGEMENT*

Ingredients

8 oz (240 ml) vodka
12 oz (360 ml) wine conditioner
12 oz (360 ml) *cuvée* (from the bottle set aside at the beginning)
½ tsp sulphite crystals

Equipment

Wide tub — bushel-basket style
2 lb (1 kg) rock salt (to mix with ice)
5 bags crushed ice
Plastic champagne corks
Champagne wires

Procedure

At this step, commercial wineries chill the bottles in the riddling board to 30°F (–1°C) to avoid losing too much CO_2 during the *dosage.* If possible, chill the bottles to this temperature, either by placing them outside (if it is winter) or by packing them with ice. Do not return the bottles to an upright position or disturb the sediment in any way.

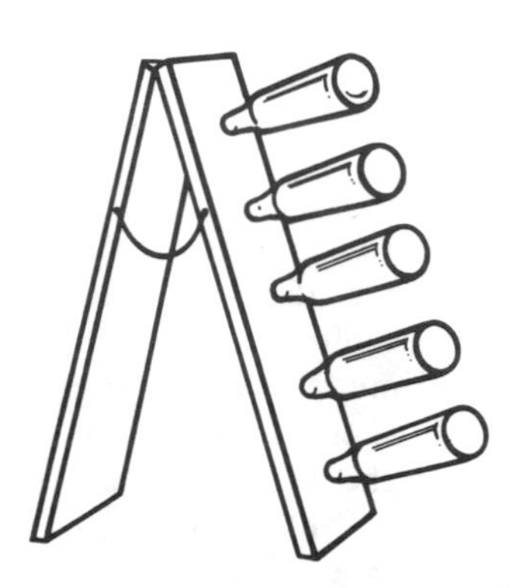

Riddling board at the beginning of the 6-week period

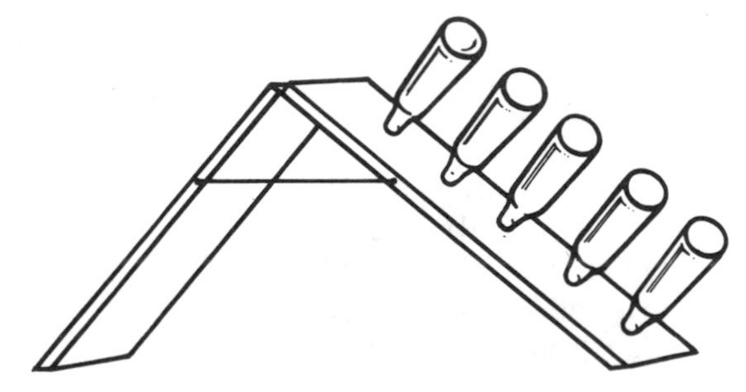

Riddling board at the end of the 6-week period

11. When the bottles are chilled, take one and open it. If there is no release of CO_2, the proper fermentation has not taken place. Do not proceed with the *dégorgement.* Return the wine to the carboy and referment.

12. If proper fermentation has taken place, place the ice and the salt in the wide tub. Carefully insert the bottles into the ice, neck down, to a depth of 6 inches (15.4 cm). Leave them in position until the sediment in the bottles has frozen into a 1-inch (2.5 cm) plug.

13. Mix all the ingredients thoroughly in a glass pitcher and chill the mixture in the refrigerator. This mixture is the *dosage.*

14. Place the primary fermentor on its side. Take a bottle, aim it into the primary fermentor, and remove the crown cap. The ice plug will blow out. (This can be a messy procedure.) Repeat with each bottle.

15. When the ice plug has blown out, fill each bottle with 1 ounce (30 ml) of *dosage*, or to within 1 inch (2.5 cm) of the top. Insert a plastic champagne cork and wire it down.

This will produce a dry *brut* champagne. Store the bottles upright in a cool, dark place. Chill well before serving.

Champagne—Bulk, or Modified Charmat, Process

The modified Charmat process requires an investment in equipment of approximately $400. But if you love champagne, it is the easiest way to make large quantities.

Equipment

20 lb CO_2 tank
Pressure gauge
Regulator gauge with setscrew to adjust the gas flow
Gas flow line with a quick-disconnect valve
Product flow line with a quick-disconnect valve
Serving tap
5 gallon (20 lit) stainless-steel product tank
Champagne bottles
Plastic champagne corks
Champagne wires

You can buy a 20 lb CO_2 tank at a soft-drink distributor or a winemaking supply store. Be sure to obtain very detailed instructions for using this equipment. There are several brands, and they have different-sized connections, which means they are not interchangeable. It is wise to obtain all your equipment from one supplier.

Ingredients

5 gal (20 lit) *cuvée* — sound, dry white wine of 10.5% alcohol and 6.5–7.0 g/lit acid. It should be 6 months old, fined, filtered, and chill-proofed.

12 oz (360 ml) wine conditioner*

**Caution:* Sweeten the *dosage* with wine conditioner only. If sugar is used without potassium sorbate, refermentation will occur and your containers will explode.

Procedure

1. Pour the wine conditioner into the product tank. Syphon the *cuvée* into the product tank. Mix thoroughly.

2. Chill the tank as near to 30°F (–1°C) as possible.

3. Charge with 45 psi carbon dioxide.

4. Rock the tank to encourage the CO_2 to go into solution. Putting the tank in the trunk of a car for a few days is very effective.

5. Run the carbonated wine into frozen champagne bottles, insert plastic champagne corks, and wire them down. Or, alternatively, decant the wine from the tank into chilled champagne bottles as needed to serve.

Champagne — Andovin Method

The Andovin method is similar to the méthode champenoise, *but much easier.*

STEP ONE

Equipment

26 oz wine bottle with plastic cap to fit
5 gallon (20 lit) carboy
Long-handled wooden or plastic stirring spoon.
Syphon hose

Ingredients

5 gal (20 lit) *cuvée* — sound, dry white wine of 10.5% alcohol and 6.5–7.0 g/lit acid. It should be 6 months old, fined, filtered, and chillproofed.
13 oz (390 g) white cane or beet sugar
5 tsp yeast nutrient
2 x 5-g pkts champagne wine yeast

Procedure

1. Syphon off 16 ounces (480 ml) of *cuvée* into the wine bottle, seal the bottle with a plastic cap, and set aside in the refrigerator to be used as *dosage*.

2. Syphon off 16 ounces (480 ml) of *cuvée* into the glass pitcher. Add the sugar and yeast nutrient; stir until both solids have dissolved.

3. Add the yeast to 1 cup of warm water (100°F [37°C]). Let it stand 10 minutes.

4. Pour the sugar and yeast nutrient into the carboy. Add the yeast.

5. Syphon the remaining *cuvée* into the carboy. Stir with the long handle of the spoon. Attach the fermentation lock.

6. Let stand 24 hours in 75°C (23°C) temperature.

STEP TWO

Equipment

25 x 25-oz (750-ml) champagne bottles
26 crown caps (good quality)
Crown capper

Procedure

7. After 24–48 hours, when the *cuvée* is cloudy with growing yeast, assemble the champagne bottles. Wash them, but do not sulphite them.

8. Remove the fermentation lock from the carboy and stir the yeast into suspension with the long handle of the spoon.

9. Syphon the *cuvée* into the champagne bottles, leaving 1 inch (2.5 cm) of air space at

the top of each bottle. Crown-cap the bottles.

10. Store the bottles in an upright position for 2 months at 65–75°F (19–23°C). Twice a week during this time, invert each bottle, shake it gently to encourage fermentation, and return it to an upright position. A substantial yeast deposit will develop in the bottles. Let rest 1 month to firm the deposit.

STEP THREE

Caution: The bottles are now under pressure and dangerous. From this point on, always wear safety gloves and eye protection when handling.

Equipment

Eye protection
Safety gloves
50 x 25-oz (750-ml) champagne bottles
25 plastic champagne corks
25 champagne wires

Ingredients

8 oz (240 ml) vodka
12 oz (360 ml) wine conditioner
12 oz (360 ml) *cuvée* (from the bottle set aside at the beginning)

11. Assemble 25 champagne bottles. Wash them, but do not sulphite them.

12. Make a *dosage* by mixing all the ingredients thoroughly in a pitcher.

13. Pour 2 tablespoons (30 ml) of *dosage* into each bottle.

14. Place the 25 bottles containing the *dosage* into a freezer. Stand them upright.

15. Place the 25 bottles containing the *cuvée* into the freezer. Stand them upright.

16. When ice has started to form on the *cuvée* (a period of 1–1½ hours), decap each bottle and pour it carefully into a bottle containing frozen *dosage*. Do not transfer the wine sediment to the *dosage* bottle; leave it at the bottom of the *cuvée* bottle. Do not remove your safety gloves; they prevent the transfer of heat to the bottles.

17. As soon as you fill each *dosage* bottle with *cuvée*, insert a cork and wire it down. Store the bottles upright in a cool place.

18. When the *dosage* thaws, mix it into the wine by gently swirling the bottle.

19. Rechill the champagne before serving.

Champagne — Dispatch Method

The dispatch method injects carbon dioxide from a liter-size injector cylinder (trade name: Roto-Gas) into plastic soft-drink bottles fitted with a pressure valve (trade name: Pet). Because of the limited capacity of the equipment, it is better to sparkle only 1 gallon of cuvée *at a time. This is the quickest method to date of sparkling wines. You can dispatch champagne into the hands of thirsty guests at a few hours' notice. You can also sparkle your rosé wines and make sodas.*

Caution: ***Never use glass bottles. Always supervise children using the equipment.***

Equipment

1-lit Roto-Gas CO_2 injector cylinder
4 x 1-lit or 2 x 2-lit Pet plastic soft-drink bottles with screw-top pressure valves
Small funnel

Ingredients

4 qt (4 lit) *cuvée* — sound, dry white wine of 10.5% alcohol and 6.5–7.0 g/lit acid. It should be 6 months old, fined, filtered, and chillproofed.
1 tsp sulphite crystals dissolved in 1 qt (1 lit) of water
2 oz (60 ml) wine conditioner

Procedure

1. Rinse the soft-drink bottles with sulphite solution.

2. Add the wine conditioner to the gallon of *cuvée*. Mix well with the long handle of the spoon.

3. Using the funnel, fill each bottle to the shoulder with *cuvée*. (This leaves room for the gas.)

4. Attach the screw-top valves to the bottles and chill them in the freezer at 30°F (–1°C) until small ice crystals form (a period of about 1 hour). Take care not to let the *cuvée* freeze solid, or you will weaken the walls of the bottles.

5. Invert each bottle over the nozzle of the CO_2 cylinder and allow gas to bubble in for 3–4 seconds. With the increased pressure, the walls of the bottle will become very firm.

6. Remove the bottle and shake it vigorously. Its walls will become soft again as the gas is absorbed by the liquid.

7. Bubble in more gas and shake the bottle until its walls become soft once more. Repeat the process until the walls of the bottle remain firm even after vigorous shaking.

8. Store the bottles in the refrigerator until use.

Note: If you do not serve all the champagne you have prepared, you can store it indefinitely in the refrigerator. Simply recarbonate each bottle before serving.

Plum Liqueur

READY: 4 MONTHS

A light-alcohol liqueur made from fresh damson plums

INGREDIENTS

1 lb	Damson plums (with the stems removed)	450 g
1½ cups	Dark brown sugar	360 ml
3 oz	Glycerine	90 ml
24 oz	Vodka	720 ml

EQUIPMENT

Sealable gallon fruit jar
Darning needle

PROCEDURE

1. Prick each plum with a darning needle several times.
2. Sterilize a fruit jar with boiling water.
3. Place all the ingredients in the jar and seal it tightly.
4. Every 3 weeks, invert the jar so that it spends 3 weeks on its base, then 3 weeks on its lid, and so on.
5. After 3 months, drain off the liqueur and bottle it. Chill before serving.

Cherry Liqueur

READY: 4 MONTHS

A light-alcohol liqueur made from fresh cherries

PRIMARY INGREDIENTS		
40 only	Fresh, ripe cherries (with the stems removed)	
6 oz	Sugar	180 g
26 oz	Rye whiskey	780 ml

EQUIPMENT
Sealable gallon fruit jar
Darning needle

PROCEDURE

1. Prick each cherry with a darning needle several times.
2. Fill the jar with cherries and add sugar.
3. Pour in whiskey.
4. Seal the jar tightly.
5. Every 3 weeks, invert the jar so that it spends 3 weeks on its base, then 3 weeks on its lid, and so on.
6. After 4 months, drain off the liqueur and bottle it. Chill before serving.

Pat Hansen's Brandied Fruit Mélange

Staff home economist on the Amateur Enologist, *which we published for 15 years, Pat Hansen came up with this recipe for "crocked fruit." It's the best we've ever tasted. Use your own liqueur or sherry, but you will have to buy the brandy.*

PRIMARY INGREDIENTS

16 oz	Brandy	480 ml
16 oz	Kirsch liqueur	480 ml
16 oz	Sherry	480 ml
6 cups	Strawberries	1.5 lit
6 cups	Sugar	1.5 lit
1 tbsp	Grated lemon rind	
1 tbsp	Grated orange rind	
In a mesh bag:		
1 tbsp	Whole cloves	
1 tbsp	Whole allspice	
2	Cinnamon sticks	
1 tbsp	Fresh gingerroot (chopped)	

EQUIPMENT

3-gallon crock with cover (in perfect condition)
Large straining bag
Small mesh bag for spices
Sulphite solution for sanitizing the crock.

PROCEDURE

1. Sanitize the crock with sulphite solution.

2. Place the brandy, kirsch, sherry, and rinds in the crock.

3. Add the mesh bag containing the spices.

4. Wash and hull the strawberries. Mash them in a large saucepan and add sugar. Bring the mixture to boil over medium heat, stirring until the sugar is dissolved. Remove the pan from heat and allow to cool.

5. Pour the mashed strawberries into a straining bag and squeeze the juice out into the crock. Discard the pulp.

6. Cover the crock and allow it to stand in a cool place.

7. As various fruits come into season, add them to the crock with an equal weight of sugar and 16 ounces of brandy. Stir well. Never add more than 2 quarts of fresh fruit (2 lit) at one time. The added fruits can be a variety or 2 quarts of one fruit. Suggested fruits: Bing cherries (pitted), apricots (halved and pitted), plums (halved and pitted), peaches (peeled, quartered and pitted), blueberries.

8. After your final addition, cover the crock tightly (put masking tape around the lid) and let it stand 3 months.

9. The following spring, if there is sufficient basic syrup mixture left, continue where you left off.

This mélange makes an elegant dessert poured over plain ice cream, cake, or puddings and a festive condiment for pork or poultry.

PART THREE

Techniques

Acidity

The three most important variables in wine that you can control are alcohol, acid, and sugar. Armed with this knowledge, you can create your own recipes.

Wines with too much acid are sour and harsh. Wines with too little acid are bitter, medicinal, insipid, and flat.

All wines contain fruit acids. Some are fixed, such as tartaric, malic, citric, and succinic acid; others are volatile, such as acetic and propionic acid. Some occur naturally in the grape, but acetic and propionic are formed during fermentation.

It is important to have the correct acid balance for the wine you are making. We can assess acid balance in two ways: by titration, which measures the weight of acids in a given volume of juice or wine (in grams per liter or parts per thousand), or by measuring the pH value. Scientists describe the pH value as the negative logarithm of the hydrogen ion concentration. From this is derived a scale of 1–14, on which 7 is a neutral point, 7–1 is a scale of increasing acidity, and 8–15 is a scale of increasing alkalinity.

Increasingly, commercial growers are calculating their harvest time by taking pH readings of their grapes, in order to maintain better wine color and flavor control. Generally speaking, when grapes mature, their pH value increases as their malic acid is replaced by tartaric acid, and a corresponding decrease in their titratable acid can be observed. But with fruits such as plums, the correlation between pH and titratable acid is by no means so clear cut. In fact, we have tested plum juice with a pH of 4 (too high for good wine) that had a titratable acid content of 12 g/l (twice what a good wine needs). This is why, in some of our recipes, we use only small amounts of fruit and insist on adding grape concentrate as a buffer.

Thus, though pH is a useful guide, and it can be measured to some degree by litmus and pH indicator papers, a simpler and more reliable guide for the home winemaker is acid titration.

ACID TITRATION

The principle of acid titration is very straightforward: You take a carefully measured quantity of juice or wine and add to it small amounts of alkali until it no longer tests positive as acidic. At this point you know all its acid is neutralized. By noting the amount of alkali you used to neutralize it, you can calculate the amount of acid originally present.

In American wineries, acid is measured on the assumption that it is all tartaric. In France it is measured as if it were sulphuric. We have chosen the American method.

As the test alkali, most enology textbooks suggest using a sodium hydroxide solution of $\frac{1}{10}$ normal strength, but this requires an extensive table to interpolate the results. We use a $\frac{1}{5}$ normal solution. It requires neither tables nor mathematics; and as long as 15 milliliters of juice is used, it gives a direct answer in grams per liter.

Procedure

When using an acid-testing kit purchased from a winemaking supply store, follow the instructions provided. As we have noted, sodium hydroxide solutions differ in strength. The following test is based on the use of $\frac{1}{5}$ normal strength solution and 15 milliliters of wine or

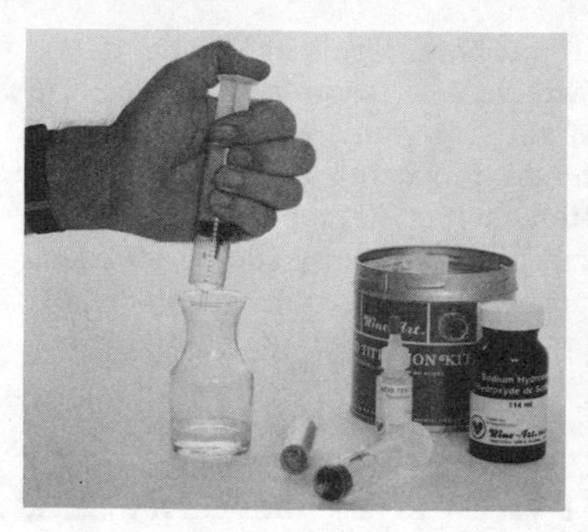
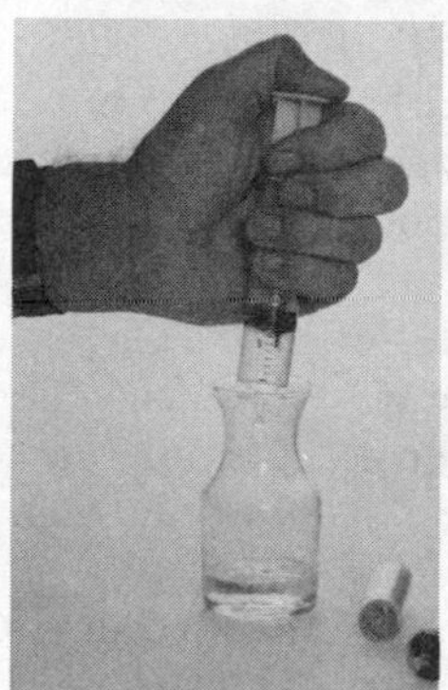
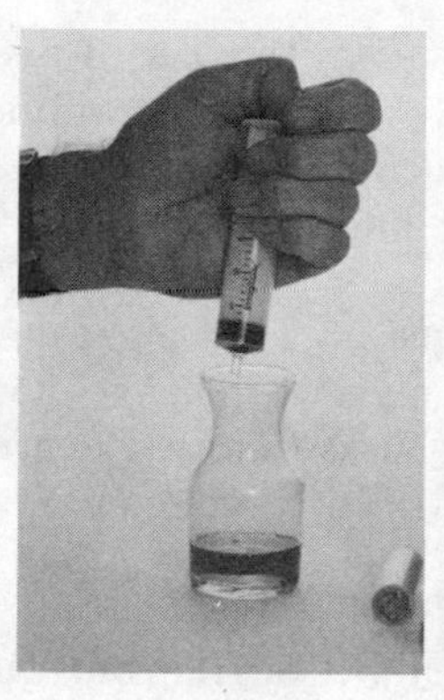
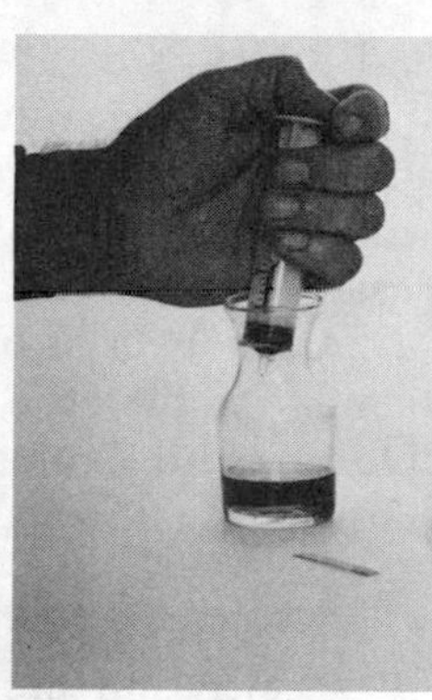

Titration

juice. Make sure all your equipment is clean and dry to start. Work in a good light source. Measure amounts accurately.

Equipment

Litmus paper
1 vial (3-oz)
1 eye dropper
2 syringes or burettes (20-ml)

Solutions

4 oz (120 ml) sodium hydroxide solution (⅕ normal strength)
1 oz (30 ml) phenolphthalein color solution

1. Using a syringe or burette, draw up 15 milliliters of the juice or wine to be tested and empty it into the vial.
2. Using the eye dropper, add 3–4 drops of color solution. Swirl it around the vial to mix it well.
3. Using the clean syringe or burette, draw up 10 milliliters of sodium hydroxide solution.
4. Using both hands to control the plunger of the syringe, release 1 milliliter of sodium hydroxide into the vial, then swirl the vial to mix. The hydroxide solution will show a tinge of pink as it hits the juice or wine, but this will vanish as you mix.
5. Repeat. After 3–4 milliliters have been added, slow down. Start adding the sodium hydroxide drop by drop. Note that now the pink color takes longer to disappear.
6. The end point is reached when the acid in the solution is neutralized. In white wines this occurs when the sample remains faintly pink through 30 seconds of swirling. Red wines or musts become a greeny black.
7. Note how many milliliters of hydroxide were needed. Each milliliter of hydroxide neutralizes 1 gram of acid as tartaric. The accuracy of this procedure is within .5 g/l of total acid.

Red Wines and Juices The end point is easy to observe in white wines and juices; however, it is more difficult to detect in reds. Litmus paper makes the end point easier to observe. When you think you are close to reaching the end point, dab a drop of the testing sample onto a piece of litmus paper. Observe any color changes.

Red indicates the solution is still acidic — the end point has not been reached.

Blue indicates the solution is alkaline — the end point has been passed.

It may be helpful to dilute deep red wines

and juices with 30–40 milliliters of distilled water. Distilled water is neutral; it will neither add nor subtract acid from your sample; but it will dilute the color and make the end point easier to observe.

An indirect or opaque white light under or behind the flask can make the change more visible. When the color turns from red to green to purple again, the end point has been reached.

Points to Remember

- Titration must be carried out quickly because the carbon dioxide in new wines, and even in the air, can interfere with the end point.
- Always recap the bottle containing your sodium hydroxide solution immediately after testing; it deteriorates rapidly when exposed to air. To ensure an accurate measure, purchase fresh solution every 6 months.
- Sodium hydroxide is caustic and a poison. Phenolphthalein has a very strong laxative effect and is also a poison. Store both out of reach of children.

USING THE RESULTS OF TITRATION

Ideally, acid levels will be as follows:

Red table wines	5.5–6.5 g/lit
White table wines	6.0–8.0 g/lit
Sherry	3.5–5.0 g/lit
Port	4.5–5.0 g/lit
Champagne	6.0–7.5 g/lit

Raising the Acid Level

Low acid is sometimes a problem with California grapes. If your total acid reading is 6.0–7.5, do nothing. If it is 5.0 g/lit or lower, bring it up to 6.5 g/lit for reds, 7.5 g/lit for whites.

Thus, if you have 5 gallons (20 lit) of red wine must with an acid reading of 4.5 g/lit, you will need to add 2.0 grams per liter (4.5 subtracted from 6.5). Since you have 20 liters, this will add up to a total of 40 grams (20 multiplied by 2). You can provide this by adding 40 grams of Vinacid, acid blend, or another commercially available acid additive. Use acid "O" for red wines; acid "R" for white wines.

When adding acid, however, caution is advisable. Commercial acid preparations differ in strength. It is easy to raise the acid level, but very difficult to lower it. It is therefore advisable to add acid in very small quantities. Dissolve 1 ounce (30 g) of acid additive in ½ cup of hot water and stir it into the must. Check the acid level with a second test before adding more.

Lowering the Acid Level

High acid is often a problem with hybrid grapes from northern and eastern parts of the U.S. Acidex is the best-known commercial product used to reduce acid levels, but food-grade calcium carbonate (chalk) is also suitable. The method of use for both additives is the same.

Calculate the amount of chalk you need as follows:

If your must contains 12 g/lit acid, and the desired level is 7 g/lit, you will need to remove 5 g/lit. If you have 100 liters of must, the total weight of acid to be removed will be 500 grams (5 multiplied by 100). Chalk will neutralize 80% of its weight; therefore, you will need to add 625 grams (500 divided by .8).

Syphon off 4 gallons (16 lit) of liquid from the must into a plastic tub. Add the chalk to it little by little. For the next 24 hours, stir the liquid every 2–3 hours. Then rack the liquid off and return it to the must, taking care to leave the sediment behind in the tub.

The must should now have an acid content of 7 g/lit. When this is chillproofed, more acid will precipitate out as potassium bitartrate, leaving an expected acid level of 5.5–6.0 g/lit — perfect for a red wine.

Note: American enologists often refer to titratable acid content as a percentage of total acids. This figure is arrived at simply by inserting a decimal point, so that 6 g/lit becomes .6%. We find grams per liter or parts per thousand a clearer expression.

ACID INCREASE DURING FERMENTATION IN WINES MADE FROM CONCENTRATES

We do not have the sophisticated laboratory equipment required to prove beyond doubt why total acids increase during fermentation in fruit wines and wines made from grape concentrates. But we believe it can be explained as follows:

Vinifera and hybrid grapes drop out large amounts of potassium bitartrate during fermentation and more later during chillproofing. This acid loss is largely offset by acids created during fermentation, such as succinic, lactic, and gluconic acid, plus the addition of 20–40 ppm sulphurous acid via sulphite. Thus, a grape must can start with an acid content of 6.5 g/lit and finish as wine with 6.0–6.3 g/lit — despite the bitartrate loss.

Grape juice concentrates, however, lose most of their potassium bitartrate during the process of concentration. Thus, even concentrates with a pH as low as 3.1–3.3 will have only 3–4 g/lit of total acid. Some concentrate manufacturers add citric acid to their concentrate, but this can be a mistake. When you ferment the reconstituted juice, you will get the usual increase in acids due to fermentation, but you won't get the dropout of bitartrates, because they are already gone. The usual acid increase in wines made from concentrate is approximately 2 g/lit. You must take account of this increase in your recipe, or you will end up with a wine too high in acid. It's much better to have to add .5–1.0 g/lit of acid to the finished wine than spoil the wine by trying to remove excess acid.

The same increase occurs in fruit wines because they contain mostly citric or malic acid, which do not drop out during fermentation like potassium bitartrate. This is why we state that a desirable acid content in wine is 6.5 g/lit, but in many of our recipes for fruit wines and wines made from grape concentrate we recommend an acid level of only 4.5 g/lit in the must.

Aging

BULK AGING

Sunlight and fluorescent light have a deleterious effect on wines. If your secondary fermentor is transparent or translucent, it is essential to bulk age your wines in darkness. The optimum temperature range is 60–70°F (15–21°C).

BOTTLE AGING

Bottle-age your wines in darkness at 40–70°F (4–21°C).

Bottling

Like all consumers, the wine drinker is influenced by packaging. If someone removed a rusty screw cap from a gallon jug labeled Vinegar and poured you some red liquid into a paper cup, you would be unlikely to enjoy it, even if it were a Mondavi Cabernet. So it makes sense to package your wines according to their worth, taking into consideration their basic ingredient, the time they need to mature, and the investment of your labor. If the bottle has an elegant label, a capsule of lead or plastic over the top, a long cork, and is served in a true wineglass, the recipient is prepared to enjoy a delicious wine created by someone who cares. First impressions do count.

Similarly, let's be practical about our quick wines. These everyday wines of 4–12 weeks' vintage hardly merit the expense and effort of a 50-cent bottle, a 20-cent label, and a 25-cent cork. The new bag-in-a-box, referred to as cask wine in many countries, is merely a cardboard box with a self-deflating plastic or foil bag inside that lives in the refrigerator, along with the milk. A plastic spout protrudes from the box; when you push up on a small lever, your cask dispenses a glass of wine or a carafeful. As you draw wine off, the bag deflates, so there is no air space to spoil the wine. One California winery is packaging superior-quality grape concentrate in these casks.* They protect the concentrate, are easier and lighter to handle than cans, and you have your cask to serve your wine in. Simply remove the rubber spout carefully, fill the bag through a small funnel, and replace the spout.

*Guimarra, Bakersfield

Good wines, fine wines, and wines of which you have great expectations and intend to age should be bottled, corked, and coded. Bottled wine matures faster than bulk wine stored in a 5-gallon carboy. Moreover, once you open a bottle of wine, you need to be able to consume the entire contents. A 750-ml bottle is just the right amount for two people to consume at an evening meal. Opening a gallon of wine, drinking two glasses, then storing it under the kitchen sink for a week will result in spoiled wine at best, and even vinegar.

First collect the bottles. Wine bottles usually contain 26 ounces (750 ml); half bottles, or splits, contain 13 ounces. A gallon of wine will therefore fill 5 x 26 oz bottles — this is where the term *fifth* came from. There are now 1-liter bottles and 1.5-liter bottles; but most food wines are still packaged in the universal 750 ml.

You can buy new wine bottles in cases of 12, for $5 a case. But you can also make a deal with the owner of a local restaurant that serves fine wines. If you collect the bottles that he would normally throw out, you can very quickly accumulate a lifetime's supply.

To prepare used bottles, remove the neck capsules, soak the bottles in a solution of Saniton PS (or another cleaning solution recommended by your winemaking store), and rinse them with clean water. Automatic dishwashers work fine, but remove the labels first, or you'll clog the exit pipe.

Procedure

When your filtered wine is ready to bottle:

1. Rinse the bottles with a double-strength sulphite solution. (See "Sulphite," page 238.)

2. Syphon the wine into the bottles to within ¾ inch (2 cm) of the base of the cork you will be using.

3. Insert the corks.

4. Allow the bottles to remain upright for 48 hours, then store them on their sides to keep the corks moist. Watch for leakers in the next few days and replace any defective corks.

5. Apply coding labels.

CODING

If you intend to assemble a wine cellar of any size at all, you will need to devise an efficient and unambiguous coding system. We are reminded of an altercation we had when one of us realized the other was marinating a tough cut of beef with our last bottle of 12-year-old Cabernet Sauvignon. Happily, the entree was delicious, and we drank slowly and with great appreciation the last half of the bottle. But we promised ourselves we would upgrade our coding system so that after 10 years in the cellar, there would be no mystery as to what each bottle contained. Marking corks is not a good practice, and we now use small white self-stick labels and keep a much more complete logbook.

CORKING

Corks

Use 1⅓-inch (34-mm) corks for wines you intend to drink within 12–18 months. Use 1¾-inch (45-mm) corks for wines to be aged longer.

Corking Machines

The purpose of a corking machine is to compress the cork small enough to insert into the neck of the bottle. Corks are expensive — up to 20 cents each — so it's important to use a machine that inserts them efficiently. There are 5 basic types available (photograph, next page).

They all work, but types A and B work better than the others. C is my third choice, because it is easier to use than types D and E.

If you are corking only 100 bottles a year, it's hard to justify a $70 machine. Purchase type C or rent type A or B from your winemaking supply store.

Procedure

1. Waxed or silicon-treated corks need no soaking. But if your corks are not waxed or silicon treated, soak them 1–2 hours in warm water containing 2 Campden tablets per quart.

2. Place a capful of vegetable oil or glycerine beside the corking machine. As you use the machine, dip your finger into the oil or glycerine after every 5 corks and lubricate the jaws or channel of the machine.

FOILS AND CAPSULES

Foils or capsules provide a very professional-looking finishing touch to bottles of wine you intend to give as gifts. Attach them just before presentation so that they are clean and fresh.

Lead foils are still the most impressive. To be rolled on firmly, they require a small device called a capsuler. Lead foils have been blamed for contributing lead to the first glass of wine poured.

Aluminum capsules are cheaper and are applied in the same way, but they are less grand than lead foils.

Enitherm capsules make a bottle tamperproof. They slide over the top of the bottle easily and, when held over the spout of steam from a boiling kettle, shrink to fit snugly.

Plastic capsules are the most popular cap-

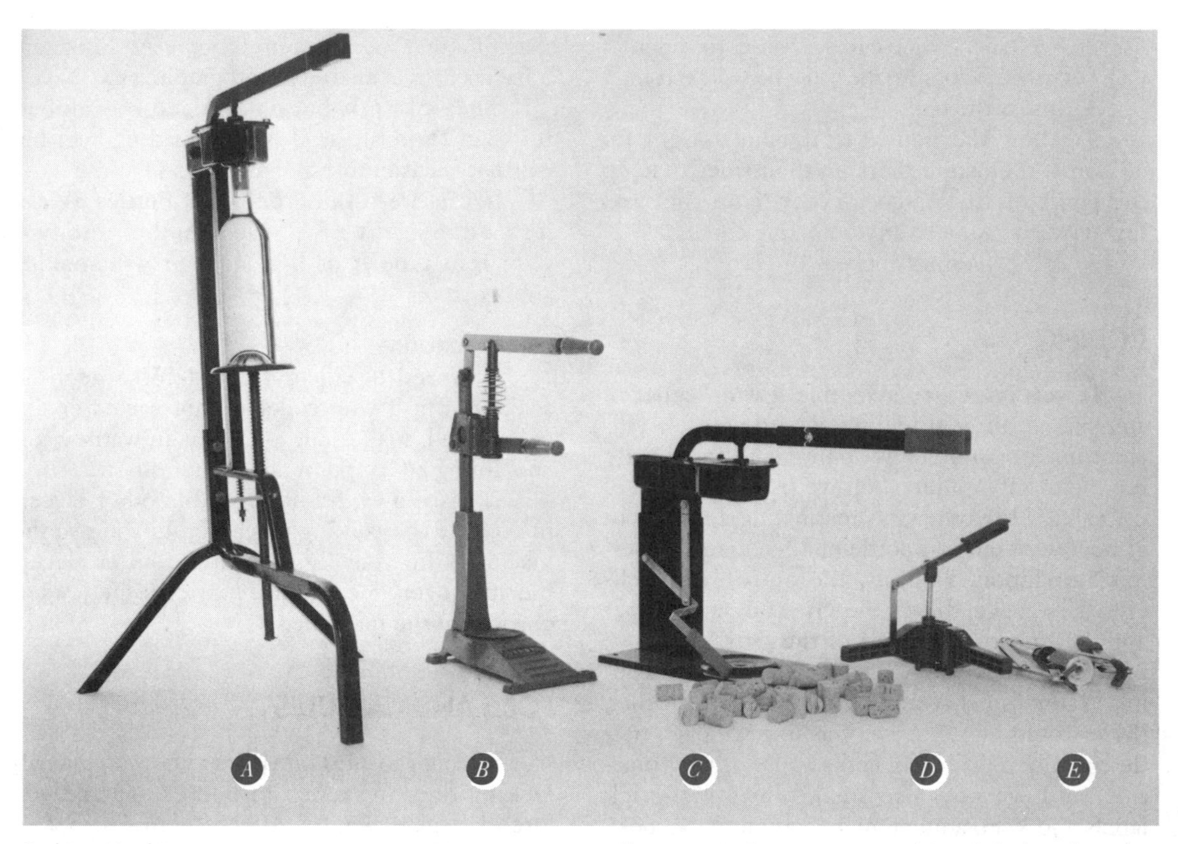

Corking Machines

sules these days. They are easily applied and fit tightly. They have a pull tab at the top, which reveals the cork for easy extraction. To apply them, you immerse them in hot tap water, slide them over the top of the bottle, and they shrink to adhere.

DECORATIVE LABELS

Though not a necessity, labeling is the final touch. Most winemakers fancy-dress their wines for special occasions. There are a variety to choose from, including simulated French and German styles, or you can have labels printed to your own design. They become dusty and discolored when stored for any length of time, so apply them just before serving the wine or presenting a bottle as a gift.

Fermentation Lock

Often called a bubbler or a breather, the fermentation lock is designed to let carbon dioxide escape from your wine during fermentation, but prevent air molds, bugs, and bacteria from entering. It is, in effect, a low-pressure valve; the gas bubbles out, but spoilage organisms cannot penetrate the sulphite solution. Three designs are commonly used:

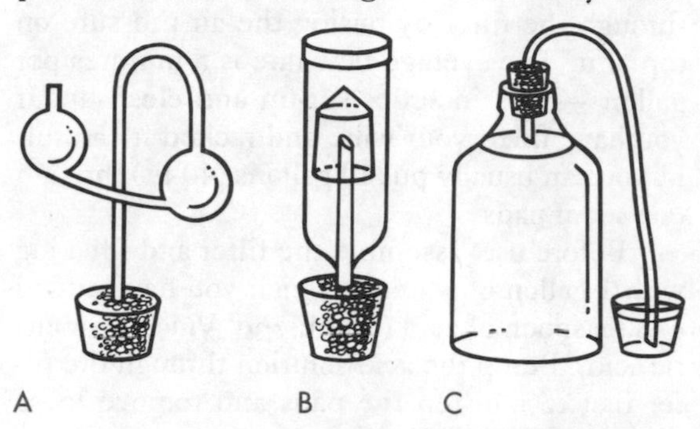

The most popular design is B. You remove the plastic cap and fill it with sulphite solution to the midway line so that the float becomes suspended in the reservoir. To achieve an accurate filling, this is best done at eye level. Replace the cap and insert the stem into a hollow rubber bung fitted to the neck of your secondary fermentor.

Fermentation lock A is also effective. The danger with C is temperature drop: If the wine in the fermentor cools the alcohol contracts, and the sulphite solution and accumulated dust will be sucked up into your wine.

Whichever design you use, always check the sulphite level every 3 weeks. In low-humidity areas, such as the desert, check every 3 days. If the solution dries up, your wine is without protection.

If you are caught without a fermentation lock in a moment of need, take a clean plastic bag 6–8 inches (15-20 cm) square and slip it over the neck of your carboy; then secure it with an elastic band. This is not the best protection, but it will suffice until you can get a proper lock.

Remember, the single largest cause of wine spoilage is air. No matter how scrupulously you maintain your fermentation lock, your wine will spoil unless you keep your carboy topped up, too.

Filtering

Eighteen years ago, when we wrote *The Art of Making Wine*, we said there was no suitable filter available to home winemakers that would do the job commercial wineries are able to do with their wines. That statement is no longer true. You can now filter hundreds of gallons of wine as effectively as most wineries.

The breakthrough came with a Danish pressure-filtering system called the Vinamat, and there are now a number of comparable brands on the market. We have tested several of these in our laboratory and can recommend the Vinamat and the Polyrad. Both these filters use commercial-grade filter pads and can be connected to pumps and CO_2 pressure tanks for larger volumes of wine. The Polyrad has the capacity to hold extra plates and pads in series — like a winery filter, only in miniature. This means you can filter large volumes without changing clogged pads.

There are still some gravity filters on the market. But they are slow, messy, and tend to oxidize the wine. They are all right for 1-gallon quantities, but our recipes are for 5 gallons or more.

When your wine is crystal clear, you might ask, Why filter? The answer is that most wineries filter *all* their wine because filtered wine is more stable and ready to drink sooner. Even fined wine has a great deal of protein matter

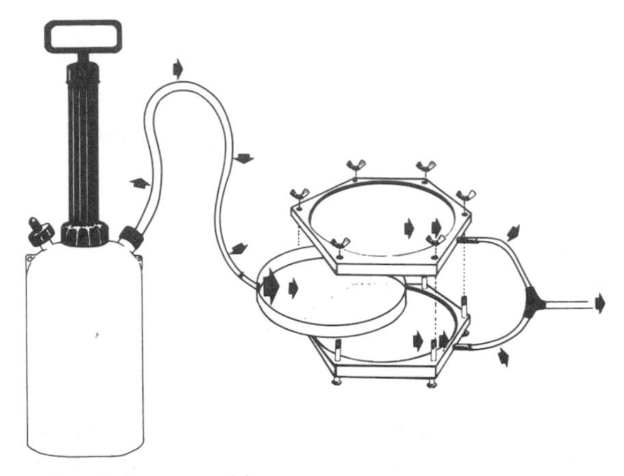

Polyrad filter assembly

suspended — up to 10,000 yeast cells per ounce (30 ml). We suggest you try a simple test: Set aside one glass of unfiltered wine, then filter the rest of the batch. Take a glass of the filtered wine, then taste and compare. We did and we now filter all our wines!

You can filter quick wines and start drinking them a day later. But with premium wines, you should filter after fining and before bulk aging. Big red wines from fresh grapes, for example, we would fine in February, filter in March with a coarse or medium filter pad, and bottle in September or October. White wine from fresh grapes, we would fine in December, chillproof in January, filter in February, and bottle in April.

Fruit wines should be fined, filtered, and bottled within 3–4 months, unless they are high-alcohol aperitif or dessert wines.

VINAMAT AND POLYRAD ASSEMBLY

Vinamat and Polyrad filters go together like sandwiches. First the base, then the pad (smooth side down), then the separator ring, then the pad (smooth side up), then the top plate. With the Polyrad, this sandwich can have several layers of filtering. (See illustration.)

The filters come with a 3-gallon (12-lit) pressure chamber and pump. You push the wine through the filter by raising the air pressure on top of it. The average flow rate is 5 minutes per gallon — this includes set-up and clean-up. If you have fined your wine and racked it carefully, you can usually put 10 gallons (40-lit) through one set of pads.

Before use, assemble the filter and fill a jug with 1 gallon of water to which you have added one teaspoon of acid (acid blend, Vinacid, or citric acid). Pump the acid solution through the filter first to moisten the pads and remove loose fibers.

Fill the tank with wine to the indicated line, then commence pumping until liquid flows. First will come water from the pads, then after a short delay, the wine. Always run the wine into a container that has been rinsed with sulphite. The trace of sulphite will remove the oxygen that the pumping has introduced into the wine.

With regular wines, allow the wine to rest at least a month before bottling.

ALTERNATE METHODS

You can connect your filter to a 5-gallon (20-lit) stainless-steel soft-drink tank and connect the tank to a carbon dioxide or nitrogen pressure tank. Only 3 psi is needed to move the wine through the filter. This method has the advantage of requiring no manual effort and no oxygen pumping into the wine. The only disadvantage is the high cost of the equipment — over $300 at the time of writing.

There is another economical alternative on the market. This is a pressure keg designed for beermakers called a Roto-Keg. It is a 6-gallon (24-lit) plastic keg that will tolerate 20–25 psi. It has fittings that permit you to connect it to a filter and use another Roto-Keg product called a Roto-Gas pressure canister. This canister is like a hairspray canister, but it can be repeatedly refilled with carbon dioxide for about 50 cents per fill. Two of these pressure canisters of CO_2 will push 6 gallons (24 lit) of wine through the filter while you stand and watch. They also protect your wine from oxygen. The cost is approximately $60.

FILTER PADS

Most suppliers carry three grades of filter pads: coarse, fine, and sterile. We generally recommend coarse for red oxidative wines, fine for reductive white wines. We seldom use the sterile grade, because it removes too much body from the wine.

Because of reports that asbestos is a health hazard, many people will have nothing to do with asbestos filters. But for 40 years, all filter pads used in wineries contained asbestos because it is a super-efficient filter medium. During the recent health scare, a study showed that in Vancouver, Canada, there were more asbestos fibers in the drinking water than in locally made wines — and Vancouver boasts the purest tap water in North America. Government health authorities in England, the U.S., and Canada tried to point out that the asbestos fibers in winery filters were not the kind that destroyed your kidneys — but to no avail. The media likes to scare you, not comfort you; so these findings received little publicity. Bowing to consumer pressure, the filter-pad suppliers provided filter pads using cellulose. We still use asbestos pads because they are more effective. When the AMA reports a rash of kidney failures as a result of drinking filtered wine, then we'll switch.

How Fine to Filter?

A micron is one millionth of a meter. The smallest object you can see with 20/20 vision is 40 microns. Blood cells are 6–7 microns, and yeast cells are as small as 1 micron. There are bacteria even smaller. This will give you some idea of the length to which you must go to sterilize wine by filtration. Sterilization is not worth the effort when the wine comes out of the filter into an environment filled with the omnipresent yeasts, molds, and bacteria we all live with.

Below are listed the degrees of filtration achieved by commercially available filter pads. The porosity is given in microns.

Filtersheet grade in mean pore diameter

AF 1	Coarse	10 microns
AF 3	Medium	4 microns
AF 4	Sterile	.8 microns

Some commercial wineries use membrane filters in the range of 1 micron, .65 micron, and .45 micron that take most of the bacteria from the wine as well. At the present time, we know of no retail outlet offering these filters to the home winemaker.

Fining

Fining is the process of clearing wine after the fermentation is complete. It removes some of the tannin, softens the wine, and greatly improves stability. You can fine and not filter, but you should not filter without fining first.

To fine a wine, you add a substance that will cause the suspended proteins and other solids to drop to the bottom. The fact that a wine appears clear does not mean fining is unnecessary. Yeast cells will be present and they will keep on settling out year after year, putting a deposit of dead yeast in the bottle and creating stability problems.

Fining was practiced by the earliest winemakers. Many substances have been used: ox blood, egg white, casein from milk, air bladders from fish, gelatin from horse hooves, seaweed — in fact, almost any protein. Clay and various types of earth have also been used.

There are now several commercial fining products on the market — organic, inorganic, and a combination of both. All have something to recommend them. The cellar master in a winery may use several types, depending on the requirement of his wines.

Generally, because every wine is different, an enologist will take samples of a wine to be fined and test them in the lab. To identical volumes of wine, he will add varying amounts of finings. For example, to one gallon he will add a gram of fining; to another he will add 2 grams; and to another, 3. He may even use 6 samples, 3 with another type of fining. He will observe which sample clears fastest, which is closest to star bright, and which has the firmest deposit. On the results of his tests he will choose the fining and the amount he will use to fine his 10,000-gallon tank of wine.

Some finings should not be used unless the wine can be later filtered with a very fine filter pad. Two in particular are the popular commercial product called Sparkolloid, and PolyClar. Sparkolloid is a combination of organic and inorganic substances. It appears to work quickly and well, but if you don't filter, it keeps dropping a light deposit for years that ends up looking like swirling smoke in the bottle.

PolyClar is a trade name for polyvinylpolypyrrolidone (PVPP). It is a synthetic compound with a highly absorptive surface. It should always be removed by filtering.

Gelatin is an organic fining that works well in red wines, especially if you use gelatin produced specifically for fining wine and beer. But it will make wine cloudy if there is not enough tannin present. With white wines, it is wise to add tannin, wait 24 hours, then add the gelatin. The best commercial brand to date is Klarospane.

Isinglass is an organic fining made from the air bladder of the sturgeon. Russian isinglass is the best, but it is expensive and difficult to prepare. It does not keep well, and it sometimes smells fishy. It too requires added tannin.

Bentonite is an aluminum silicate clay. The best for fining wines comes from Wyoming. Bentonite is very popular among California winemakers for most of their table wines, and it works well without added tannin. It is an alternative when all else fails. For instance, if your gelatin fining failed and left your wine cloudy, you could solve the problem with bentonite. German and French enologists, however, have reservations about bentonite for their best table wines; they think it affects the flavor.

PROCEDURE

First, take a hydrometer reading to make sure fermentation is complete. If in doubt, perform a sugar test. (See "Residual Sugar," page 272.) Alternatively, fill a bottle with wine, put a cap or cork on it, and place it in the warmest part of the house for 3–4 days. If, when you open it, no foam or bubbles are visible, the fermentation is complete.

Wineries chillproof their wine first, storing it at 24–30°F (–4– –1°C) for 3 weeks to precipitate any excess bitartrate crystals. But unless you have a large refrigerator kept for chilling wines only, or the outside temperature will run at 28–35°F (–2–2°C) for 2–3 weeks, you will have to skip chillproofing.

The next step is to determine which fining agent will clear your wine best and quickest. If you have a large amount of wine, it is a good idea to take samples of several fining agents and perform a practical test on a small amount first. Twenty-five ounces is a sufficient test quantity.

With gelatin, the manufacturer usually recommends the ***minimum*** amount, so you may need more than the package indicates. Fining with a bentonite-gelatin combination is quite common; the bentonite compacts the fluffy gelatin for easier racking and filtration.

Fining with Gelatin — Red Wines

To fine 5 gallons (20 lit) of wine:

1. Soak 1 teaspoon of gelatin fining in ½ cup of cold water for 1 hour.
2. Stir or whip vigorously.
3. Heat the mixture and bring it to a boil.
4. Remove the mixture from the heat and stir it into the wine.
5. Leave the wine to stand for 10–15 days.
6. Rack the wine off the sediment into a clean, sulphited carboy.
7. Top up with cold tap water and attach a fermentation lock.
8. Leave to mature 3–9 months before bottling.

Gelatin Fining — White Wines

Because the action of gelatin is based on polarity (gelatin has a positive charge and attracts negatively charged particles, such as grape tannin), it is essential that there be enough tannin in the white wine for the fining process to work. Add ¼ teaspoon of tannin per gallon (4 lit) of wine 24 hours before you add the gelatin finings.

Follow the instructions for fining red wines with gelatin, but allow 10–18 days to pass before racking the wine off the sediment. White wines should be bottled 1 month after fining.

Note: Red wines usually have enough tannin, but some red wines from grape concentrate need additional tannin to help fining and increase astringency. The exact quantity you need will be suggested in each of our recipes using grape concentrate.

Fining with Bentonite

If you buy bentonite from a winemaking supply store, it will have instructions for use; but in case you need them, instructions follow.

To fine 5 gallons (20 lit) of wine:

1. Place 7⁄10 ounce (21 g) (approximately 4½ teaspoons) of bentonite in a blender, together with 1 cup of warm tap water.
2. Blend for 3 minutes at medium speed.
3. Cover and let sit for 24 hours.
4. Blend again for 3 minutes.
5. Stir the bentonite into the wine and leave for 10 days.
6. Rack the wine off the sediment into a clean, sulphited carboy. Rack carefully because bentonite does not form a firm deposit.

Fining with Claro K. C.

Claro K. C. is the trade name of one of our favorite types of fining. It combines organic and inorganic substances — gelatin and silica soda. It is easy to use and renders a firm deposit. You do not have to filter to remove this product from your wine. In white wines, add tannin 24 hours prior to fining. Rack the wine off the sediment after 10 days. Instructions come with the product.

Racking

Racking means using a syphon hose to move wine from one container to another. Its purpose is twofold: It protects the wine from the air, and it leaves the sediment undisturbed at the bottom of the container. Obviously, this would not be possible if you were to pour wine from one container to another through a funnel. Wines left on their sediment too long develop off odors and flavors. It is therefore important to rack at the given time, according to your recipe.

PROCEDURES

Place the full container on a chair and the empty container below it, on the floor. Put the rigid tip of the syphon hose into your wine, near the bottom of the container but above the sediment. Bend down and suck on the end of the hose until the wine starts to flow. You can spit out or swallow the wine — new wine tastes bad, but one or two mouthfuls won't hurt you.

Another method is to fill both ends of the syphon hose with liquid while keeping them at the same height. You then clamp the outlet closed and lower it into the empty container. When you release the clamp, the liquid will start to flow.

A third method, if you are moving wine from containers with neck openings, is to place both ends of the syphon hose in their respective containers, one below the other, then cup your hands around the neck of the full container and blow into it. The increased air pressure will force the liquid down the hose. You can purchase syphon hoses with small bellows that make starting the flow easy. If you have not syphoned anything before, practice with water first.

You can always stop the process of syphoning by pinching the hose firmly between your thumb and forefinger or by using a specially designed clip that pinches the hose tightly. As long as you keep the syphon hose in a downward position, below the full container, you can move it from one receiving container to another without losing your gravity flow and having to start anew.

Specific Gravity

MEASURING SPECIFIC GRAVITY

When we measure the specific gravity of a liquid, we are measuring its density; and the density of a wine must is something the winemaker very much needs to know.

Sugar solution is denser than water — the more sugar it contains, the denser it is. Alcohol, on the other hand, is less dense than water — the more alcohol in a wine, the less dense it is. Thus, when we measure the specific gravity of an unfermented must, we are finding out how much sugar is present. When we measure the specific gravity of a fermenting must, we are finding out how far the yeast has progressed in converting the sugar to alcohol. When we measure the specific gravity of a finished dry wine, we are finding out how much alcohol it finally contains.

Specific gravity takes as its unit of measurement the density of water. Thus, alcohol, which is only .8730 the density of water, has a specific gravity of .8730. Similarly, a strong sugar solution that is 1.25 times the density of water will have a specific gravity of 1.25.

To measure specific gravity, we use a hydrometer. This is actually a very simple device that operates according to Archimedes' principle: The level at which an object floats depends upon the density of the liquid it is floating in. Thus, a hydrometer is essentially a float marked with a graduated scale: The higher it sits in a liquid, the denser that liquid is. Generally, a hydrometer scale is marked in increments of specific gravity, and this is the scale we use; but there are also other ways of expressing the density of wine.

The word *hydrometer* means "water measurer," but another common name for the instrument is *sacchrometer*, or "sugar measurer." Commercial winemakers have created additional hydrometer scales that reflect their practical interest in sugar and alcohol content. The Brix and Balling scales gauge a liquid in terms of its percentage of sugar, or soluble solids. Another scale used by French winemakers, the Baume 10 scale, measures sugar, but expresses it in terms of the potential alcohol that will result when fermentation is complete. In our recipes we give hydrometer measurements in terms of specific gravity, but we have included some conversion tables for your convenience.

If you buy a hydrometer in a winemaker's supply store, it will probably be a triple-scale hydrometer that we devised in 1960 to show specific gravity, Balling, and potential alcohol by volume all at the same time. You usually get a set of instructions with the instrument, but in case you do not, proceed as follows:

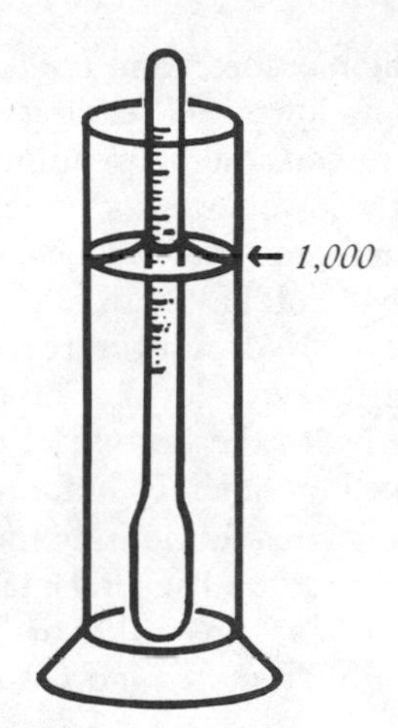

Hydrometer in a testing jar

Using a Hydrometer

1. Make sure your hydrometer is clean; even oil from your skin can affect the accuracy of your reading.

2. Using a kitchen baster or a wine thief, remove some of the wine or must that you wish to test and place it in a testing jar. Try not to include the solids — stems, seeds, or bits of fruit. One way to avoid solids is to strain the liquid through a cheesecloth first. Fill the testing jar to within an inch of the top.

3. Check the temperature of the liquid. Hydrometers are designed to be used at a specific temperature. The usual range is 59–68°F (15–18°C), but unless your test sample is more than 18°F (10°C) above or below this range, you can safely ignore the temperature.

At extremes, however, a discrepancy will

enter into the measurement. High temperatures will give you a lower SG reading; low temperatures will give you a higher reading.

4. Gently lower the hydrometer, bulb end first, into the liquid until it floats. Give it a gentle spin to shake off any bubbles.

5. Let the hydrometer rest 30 seconds, then inspect it at eye level. The reading is the number level with the bottom of the meniscus. Write it down immediately, before you forget it.

Caution: With a fermenting must, do not let the hydrometer rest 30 seconds; take your reading as quickly as possible, or bubbles will adhere to the hydrometer and raise it up.

Common Mistakes When Using a Hydrometer

1. Failing to remove the packing case before use (yes, it happens).
2. Trying to use the hydrometer upside down.
3. Dropping the hydrometer in the testing jar and breaking it.
4. Holding the hydrometer by the stem and shaking it. (This will cause the stem to snap off.)

If you pay less than $15 for your hydrometer, it will probably be guaranteed accurate only to 0.002 SG, which is adequate for home winemaking. However, you can easily check its accuracy by placing it in water within the prescribed temperature range. It should give a specific gravity reading of 1.000.

ADJUSTING SPECIFIC GRAVITY

At the start of fermentation, a specific gravity reading will tell you how much sugar you have in your must. Where it is practical, our recipes give you an ideal start figure; however, yours may be different. This could be because the sugar content of your basic ingredient is unusually high or low; or it may be because you have made an error in measuring out your ingredients. By knowing the starting specific gravity, you can make good any deficiency in the ingredients and save the wine.

If your must gives a hydrometer reading lower than the recipe calls for, you can add sugar. If it gives a higher reading, you can dilute it with water. Note, however, that a low SG reading will be obtained if your sugar has not completely dissolved. Obviously, in this case you should not add more sugar. Always stir your must carefully to make sure no sugar grains can be felt on the bottom of the fermentor before attempting to adjust your SG upward.

Calculating Additional Sugar

You can calculate the extra sugar you need quite easily. Seventeen ounces (510 g) of sugar added to 5 gallons (20 lit) of water will raise the specific gravity by .010. Thus, if your must is showing an SG of 1.050, and the recipe gives a starting SG as 1.100, you need to add 5 lb (2.3 kg) of sugar.

Note: If you are using corn sugar or dextrose, increase the amount of sugar you add by 20%.

Calculating Additional Water

Two quarts of water added to a 5-gallon must will reduce the specific gravity by .010, or 10%.

Playing Safe

If your SG is too low, you can easily add more sugar; if it is too high, you can add water. But if we have a choice, we would rather add extra sugar than extra water, because when you add extra water, you dilute the other ingredients. It is a good idea, therefore, to hold back 10% of the sugar and take a careful SG reading. If it is lower than the recommended starting SG, you simply add more until you obtain the SG your recipe requires.

ESTIMATES OF ALCOHOL YIELD

The more books you read, the more variation you will find among the estimates of alcohol yielded by given specific gravity readings. For the most part, these estimates are correct; that is, X amount of cane or beet sugar, fermented under ideal conditions, will produce X amount of alcohol. However, 25 years of careful observations have produced some empirical knowledge. Temperature, the availability of oxygen, and the strain of yeast will all affect the alcohol yield. More important, your hydrometer measures not only sugar, but all the soluble solids that are in solution with it — acids, pectins, suspended fruit solids, and so on. We can demonstrate this quite simply.

The hydrometer we designed predicts alcohol yield on the basis of yields obtained from fresh-squeezed vinifera grape juice. For example, it expresses an SG of 1.090 as 12.2% potential alcohol by volume. However, if you carefully settle or centrifuge your juice, an SG of 1.090 will yield 12.5% alcohol by volume. Similarly, a filtered or fruit must, to which water and sugar have been added to reduce acid or excessive flavor, will show an alcohol yield of at least 1% more than an undiluted must. This is because the specific gravity reading will not have been boosted by so many nonsugar suspended solids.

Temperature Equivalents

a) To convert Fahrenheit to Centigrade

F° – 32 ÷ 1.8 = C°

Example: to convert 68°F to Centigrade

68 – 32 = 36
36 ÷ 1.8 = 20
68°F = 20°C

b) To convert Centigrade to Fahrenheit

C° x 1.8 + 32 = F°

Example: to convert 15°C to Fahrenheit

15 x 1.8 = 27
27 + 32 = 59
15°C = 59°F

Conversion Table: Specific Gravity at 60°F to Balling

Assuming SG of water at 60°F is unity

Degrees Balling	*Specific gravity*	*Degrees Balling*	*Specific gravity*	*Degrees Balling*	*Specific gravity*
0.00	1.000	10.0	1.039	20.0	1.081
0.50	1.002	10.5	1.041	20.5	1.084
1.00	1.004	11.0	1.043	21.0	1.086
1.50	1.006	11.5	1.045	21.5	1.088
2.00	1.008	12.0	1.048	22.0	1.090
2.50	1.010	12.5	1.050	22.5	1.093
3.00	1.012	13.0	1.052	23.0	1.095
3.50	1.014	13.5	1.054	23.5	1.097
4.00	1.016	14.0	1.056	24.0	1.099
4.50	1.017	14.5	1.058	24.5	1.102
5.00	1.019	15.0	1.059	25.0	1.104
5.50	1.021	15.5	1.062	25.5	1.106
6.00	1.023	16.0	1.064	26.0	1.109
6.50	1.025	16.5	1.066	26.5	1.111

Conversion Table: Specific Gravity at 60°F to Balling *(Continued)*

Assuming SG of water at 60°F is unity

Degrees Balling	*Specific gravity*	*Degrees Balling*	*Specific gravity*	*Degrees Balling*	*Specific gravity*
7.00	1.027	17.0	1.068	27.0	1.113
7.50	1.029	17.5	1.070	27.5	1.116
8.00	1.031	18.0	1.072	28.0	1.118
8.50	1.033	18.5	1.075	28.5	1.120
9.00	1.035	19.0	1.077	29.0	1.123
9.50	1.037	19.5	1.079	29.5	1.125
				30.0	1.127

Specific Gravity/Potential Alcohol Tables

This table shows the alcohol yield you may expect in relation to the specific gravity and sugar content of the must.

		Weight of sugar to be added to 1 gallon of must				
Specific gravity	*Balling*	*lb*	*oz*	*grams* (per lit)	*grams* (per U.S. gal)	*Potential alcohol by volume*
1.000	0	0	0			0
1.005	0	0	1.7	13	48	0
1.010	2.4	0	3.5	26	99	0.9
1.015	4	0	5.2	37	147	1.6
1.020	5	0	7	52	198	2.3
1.025	6.5	0	9	67	255	3.0
1.030	7.5	0	11	82	312	3.7
1.035	9	0	12.7	95	360	4.4
1.040	10	0	14.7	110	417	5.1
1.045	11.5	1	0.6	124	471	5.8
1.050	12.5	1	2.7	140	530	6.5
1.055	14.0	1	4.7	181	587	7.2
1.060	15.0	1	7	172	652	7.8
1.065	16.5	1	9	187	709	8.6
1.070	17.5	1	11	201	765	9.2
1.075	18.5	1	13.4	219	833	10.4
1.080	20.0	1	15.6	236	896	11.2
1.085	21.0	2	2	239	907	11.9
1.090	22.0	2	4.3	243	1029	12.6
1.095	23.0	2	6.3	286	1085	13.4
1.100	24.0	2	9	306	1162	14
1.105	25	2	11.7	326	1239	14.9
1.110	26.5	2	14.3	346	1313	15
1.115	27.5	3	0.8	358	1361	16.4

Specific Gravity/Potential Alcohol Tables *(Continued)*

This table shows the alcohol yield you may expect in relation to the specific gravity and sugar content of the must.

Specific gravity	*Balling*	*Weight of sugar to be added to 1 gallon of must* *lb*	*oz*	*grams (per lit)*	*grams (per U.S. gal)*	*Potential alcohol by volume*
1.120	28.5	3	3.5	388	1474	16.9
1.125	29.5	3	6.3	405	1539	17.6
1.130	30.5	3	9	425	1616	18
1.135	32	3	12	448	1701	18.9

Proof Conversion Table

Note that proof is an arbitrary level of alcoholic strength, differing from one country to another. This table shows the percentage equivalents of three commonly used proof scales.

% Absolute alcohol by volume	*Canadian proof rating*	*Degrees of proof, Sykes scale*	*Equivalent U.S. proof rating*
100	75 Over proof	175	200
97	70 OP	170	194
94	65 OP	165	188
91	60 OP	160	182
86	50 OP	150	172
80	40 OP	140	160
74	30 OP	130	148
69	20 OP	120	138
63	10 OP	110	126
57.1	Proof	100	114.2
51	10 Under proof	90	102
46	20 UP	80	92
42.5	25 UP	75	85
40	30 UP	70	80
34	40 UP	60	68
29	50 UP	50	58
23	60 UP	40	46
17	70 UP	30	34
11	80 UP	20	22
6	90 UP	10	12
0	100 UP	0	0

Sulphite

Sulphite is the name commonly given to potassium or sodium metabisulphite. The home winemaker has four uses for sulphite:

1. To sanitize equipment and containers before use
2. To add to wine as a bactericide and antioxidant
3. To fill fermentation locks
4. To sprinkle over chopped fruit to prevent browning

1. SULPHITE AS A SANITIZER

In the past, the standard solution for sanitizing was 2 ounces of sulphite crystals to 1 imperial gallon of water. But currently, gallon jugs are being replaced by 4-liter jugs in Canada and the U.S. Thus, it is more useful now to think of it as 12.5 grams per liter, or 50 grams per 4-liter jug.

2. SULPHITE AS A BACTERICIDE AND ANTIOXIDANT

We specify in each recipe exactly how much sulphite to use. However, as a general principle, a fermenting must requires approximately 80–100 ppm. This means that a 5-gallon recipe will require the addition of ⅔ teaspoon (4 g) of sulphite crystals or 8 Campden tablets. When adding sulphite crystals or Campden tablets to wine, remember always to dissolve them first in a small amount of warm water. Campden tablets should be crushed first.

Here is an easy reference for adding 100 ppm of sulphite to bulk recipes:

To 100 pounds of crushed grapes, add ¼ ounce (7.5 g).

To 420 pounds (12 lugs, or boxes) of crushed grapes, add 1 ounce (30 g).

To 1 ton of crushed grapes, add 5.5 ounces (165 g).

To 50 gallons of wine or juice, add ½ ounce (15 g).

Sulphur Dioxide Loss

After the initial high dose of 80–100 ppm, to kill wild yeasts prior to fermentation, a constant level of 30–50 ppm sulphur dioxide must be maintained. Whenever sulphited wine is moved from one container to another, small quantities of SO_2 are given off as gas. To replenish the sulphur dioxide level, therefore, sulphite crystals must be added to each carboy after the final racking and before bulk aging. We specify in each recipe exactly how much sulphite to use. However, as a general principle, a 5-gallon carboy of wine will require ¼ teaspoon (1.5 g) of sulphite crystals. When adding sulphite crystals to wine, remember always to dissolve them first in a small amount of warm water.

When transferring fermented wine to bottles, it is easier just to rinse the bottles with sulphite solution. The residue in each bottle is sufficient to restore the sulphur dioxide level. Traditional wisdom suggested using a standard sulphite solution (2 ounces to 1 gallon). However, in our lab we found that this procedure added only 17–18 ppm to the wine — about half the minimum required. Therefore, we recommend that you use a double-strength solution for this purpose: 4 level teaspoons (25 g) per liter of water. Clearly label this as double-strength solution after use, to avoid confusion. Also, avoid inhaling the fumes — especially if you have respiratory problems — or they'll take your breath away.

Too Much Sulphite

With too much sulphite, a must will not ferment. If you have added too much sulphite to your must, proceed as follows:

a) Add a quantity of hydrogen peroxide to the must equal to the quantity of sulphite you added. If you added 1 tablespoon instead of 1 teaspoon, add 1 tablespoon (15 ml) of 3% hydrogen peroxide solution. Stir well.

b) Heat the must to 93°F (33°C) by applying a heating pad or by immersing sealed containers of hot water in the must.

c) When the must reaches 93°F (33°C), stir it vigorously to drive off the excess sulphur dioxide. Continue to stir as the must cools.

d) When the must cools to 70°F (21°C), remove ½ cup and pour it into a small bowl. Add ½ cup of water. Sprinkle a 5-gram packet of yeast into the liquid and allow it to rehydrate for 1 hour. (If it is available, use yeast with a high alcohol tolerance.)

e) Stir the yeast mixture into the wine.

Note: This wine will not be of premium quality because often some of the excess sulphite becomes sulphate, which imparts a bitter taste.

3. SULPHITE AS A BARRIER IN FERMENTATION LOCKS

Fermentation locks require a standard solution of 12.5 grams per liter.

4. SULPHITE TO PREVENT THE BROWNING OF CHOPPED FRUIT

Browning occurs because of oxidation. Sulphite is an effective antioxidant. To prevent apples and pears from browning while you are chopping them, mix 1 teaspoon of sulphite crystals in ½ liter of water and sprinkle this over the chopped fruit as you proceed. No additional sulphite will be needed in the recipe until fermentation is complete.

MEASURING SULPHITE IN SOLUTION

Wine needs to be protected by a minimum sulphur dioxide level at all times. If you follow a good recipe carefully, you can achieve excellent results without ever measuring the sulphur dioxide level of your wine. However, sulphite in crystal or powder form loses its strength when exposed to air. This means that even if you use the recommended quantity of sulphite crystals, your wine could still be seriously short of the required amount of SO_2 if your sulphite is old or has been poorly packaged.

Previous writers have avoided the subject of sulphur dioxide measurement for home winemakers because the procedure has been too complex to perform reliably outside a well-equipped laboratory. But in 1987 a Canadian company, Hazlemere Research, marketed an easy-to-use home testing kit under the trade name Sulfikit.* It is not only simple and convenient, it is reasonably priced at around $20, and refills for the chemicals are available. It is the only way for the serious winemaker to measure sulphur dioxide levels accurately before, during, and after bottling. The kit is accurate with white, blush, red, and rosé wines and white grape juice, but somewhat less accurate with fresh red juice.

Note: The wine industry generally measures SO_2 in ppm (parts per million). Don't be confused by the terminology. Simply take a percentage and add on four more zeroes and you have parts per million.

*See "Buyer's Guide," page 281.

Temperature Adjustment

TO RAISE THE TEMPERATURE IN THE PRIMARY OR SECONDARY FERMENTOR

Unless otherwise stated, the temperature of the must in the primary fermentor should be 75°F (23°C). To raise the temperature, place the fermentor on a heating pad or apply a heating belt. Keep a floating thermometer in the primary fermentor and monitor the increase closely. To raise the temperature of the secondary fermentor, place it on a heating pad or apply a heating belt.

TO LOWER THE TEMPERATURE IN THE PRIMARY FERMENTOR

When the temperature of the must in the primary fermentor is too high — above 80°F (27°C) — put a heavy object in a large plastic freezer bag, add ice cubes, and tie the bag tightly at the top. Lower the bag into the fermentor. Check the temperature with a floating thermometer; when it drops to 75°F (23°C), remove the bag of ice.

TO LOWER THE TEMPERATURE IN THE SECONDARY FERMENTOR

Unless the recipe calls for a warm secondary fermentation, it should be carried out at a temperature of 65°F (18°C). If a carboy is too warm, simply move it to a cooler location.

If possible, keep fermentors on a chair or bench, off the floor and away from windows.

Yeast Starters

DRIED YEAST

If you are using dried yeast, little preparation is required. In fact, if you are in a hurry, you can simply cut the sachet open and sprinkle the yeast on top of the must. Better, however, is to rehydrate it first in a cup of warm water (100–110°F [37–43°C]) for 10 minutes, then stir it into the must.

A 5-gram package is sufficient for 6 gallons (24 lit). If you need to stretch a packet for use with 12 gallons (48 lit), add a teaspoon of sugar to the warm water, cover the cup with plastic wrap, and let the yeast grow for 2 hours before use.

YEAST IN VIALS AND TEST TUBES

If you are using yeast in vials or test tubes, you will need to make a yeast starter first, to allow the yeast to increase in volume before you add it to your must. Your starter should have a volume of 1–3% of the total volume of the must. To make a yeast starter for a must of 5–10 gallons (20–40 lit), proceed as follows:

Equipment Needed	Ingredients
Sterilized 1 qt jar	1 cup orange juice (fresh or frozen)
Measuring cup	2 cups water
Measuring spoons	1 tbsp sugar
Plastic wrap	
Elastic band	

1. Preheat a canning jar.
2. Pour 1 cup of orange juice into a pan. Add to it 1 tablespoon of sugar. Bring the mixture to a boil.
3. Pour the boiling orange juice into the preheated jar. As a precaution against the jar breaking, do this in a sink.
4. Seal the jar with plastic wrap to shut out bacteria and let it cool to room temperature (70–80°F [21–26°C]).
5. Shake the vial of yeast and add it to the orange juice mixture.
6. Replace the plastic wrap quickly and place the jar in a warm place (75–85°F [23–27°C]).
7. After 2–4 days, check the jar for signs of fermentation. If the mixture is fermenting, bubbles should rise when you swirl the jar gently, and foam should appear on the surface. The starter is now ready for use.
8. If you are unable to use the starter when it is ready, feed the solution with a tablespoon of sugar and store it in a cool place until you are ready.

Saving Your Yeast

Many experts have suggested that you can preserve yeast cultures for long periods in bottles in your refrigerator. We recommend this only if you live in a part of the world where fresh wine yeast is not available, such as Japan and the Middle East. Preserve a yeast culture as follows:

1. Save the deposit left in your carboy after the third racking and add a little sterile sugar syrup.
2. Cover the jar with plastic wrap and an elastic band to exclude air.
3. Store it in your refrigerator. Wine yeast has been known to grow at 28°F (–2°C); so it may ferment in your refrigerator at 38°F (2°C).

We repeat: Best results will be obtained by using fresh yeast with every batch.

PART FOUR

Reference

Barrels

The best red wines in the world spend 1–2 years of their lives in oak barrels. The serious winemaker must therefore give some consideration to this form of aging.

SIZES AND TERMS

Originally, the word *barrel* meant a beer barrel that held 36 imperial gallons. In America, we refer to a bourbon barrel that holds 48 U.S. gallons of whiskey. Barrels that hold wines have distinct names and sizes. There are butts, pipes, hogsheads, quarters, and octaves. Some Bordeaux wines are shipped in tuns holding 200 imperial gallons.* Butts and pipes vary between 92 and 116 imperial gallons; quarters, 23–29 imperial gallons; while octaves are used specifically for madeira, port, and sherry. The barrels of wine you see in pictures and on winery tours are generally hogsheads of 55 U.S. gallons or 209 liters, this being considered the ideal size for aging wines.

Many beginners want to use 5- and 10-gallon barrels, but at such small capacities there is too much barrel surface in relation to the volume of wine. In fact, you really shouldn't bother with barrels smaller than a quarter (23 gallons).

*Nowadays it is common for European wineries to ship finished, inexpensive wines in tankers containing 2,000 liters or more.

WOODS

Barrels can be made of many types of wood, but it is generally conceded that oak is best for wine. And not just any oak: Kentucky oak is prized for bourbon whiskey, Ohio oak is preferred for Spanish sherry, while Limousin oak is considered best for cognac. Suitable oak also comes from Yugoslavia, Russia, and Spain.

The French winemakers are reputed to use only new barrels to age table wines, but even the best oak in pipes or hogsheads imparts too much oak taste to the wine the first time the barrel is used. Besides, with new barrels now costing near $300 each, it would be too expensive to discard them after one use.

Kentucky oak is processed in a sawmill, then the staves are kiln dried, which creates a concentration of vanillin aldehyde and phenols in the wood. The staves are then charred. This creates the distinctive flavor of American bourbon. In Canadian or Scotch whiskey such a flavor would be considered a fault, and wines stored in these barrels would be undrinkable.

Many years ago U.S. coopers were a very strong lobby in Washington, and they obtained legislation that compelled all distilleries to use new barrels for each batch of whiskey produced. This caused a plethora of 48-gallon charred-oak barrels that had seen only 4–8 years of use — during which time most of the harsh flavor had been nicely leached out. Ironically, this law led to the creation of a smooth Canadian whiskey and very rich distillers, as Canadian distilleries were able to purchase these used barrels for $2, plus freight.

In contrast, French Limousin oak is cut and quartered, spread out in the forest, and left to ripen in the sun and the rain for 2–3 years, during which time it is turned and rotated often. When it is mature, it is cut into staves almost twice the thickness of whiskey-barrel staves and bound with wooden hoops. Limousin oak barrels are truly a joy to behold, but they are six times the cost of the best barrels made from American oak. If you can't obtain or afford French oak, then use rebuilt American barrels.

WHAT BARRELS DO

Whether you use French oak or decharred rebuilt American oak, a barrel will impart the flavor and aroma of oak tannin to your wine. It will also allow the wine to breathe. Small amounts of water and alcohol will evaporate through the walls of the barrel, and small amounts of oxygen will enter the wine. How fast and how beneficially this process occurs will depend on the temperature and humidity of your area. High humidity means you will lose alcohol; low humidity means you will lose water. The warmer it is, the faster the process; therefore the ideal storage temperature is 57°F (14°C), which is why caves are so popular. Higher amounts of oxygen are picked up by the wine at racking, transferring, and topping up than by breathing through the wood.

TREATMENT FOR BARRELS

Unless you buy a barrel from a personal friend who has just taken his excellent wine out of it yesterday, treat any used barrel as follows:

1. Smell the barrel; if you detect a vinegary odor, use the barrel to collect rainwater or make it into a planter. You cannot get vinegar or mold out of wood.
2. If there is no vinegary odor, fill the barrel with a solution of Barolkleen (or a similar product) mixed to the strength of 1 pound (450 g) to 5 gallons (20 lit) of hot water. Leave for 24–48 hours.
3. Drain the barrel and flush it with clean water until the flushed water runs clear.
4. Mix 8 ounces (240 g) of sodium metabisulphite in 1 gallon (4 lit) of warm water, add 1 ounce (30 g) of citric acid (Vinacid will do), and pour this mixture into the barrel. Close the barrel and roll it around so that the sulphite solution touches all the interior. This will neutralize any alkali remaining in the barrel.
5. Drain the sulphite solution out and rinse the barrel well.
6. Rack in your new wine.

FERMENTING IN BARRELS

If your new wine is still fermenting when you fill the barrel, leave a few inches between the wine and the bung hole, so it won't bubble out through your fermentation lock. Cider makers are said to encourage their barrels to foam out through the bung — something to do with "letting the bad stuff escape." But we don't like the mess on the floor or the millions of fruit flies. If you start with SO_2, the bad stuff (air) is eliminated.

After the first 10 days or so, when the first vigorous fermentation is over, your barrel must remain completely filled for as long as there is any wine in it. If it is a 40-gallon barrel, make sure you have 45 gallons of wine to start. Store the extra in a 5-gallon carboy or 5 x 1-gallon jugs and use it to keep the barrel topped up. We like to do a first racking from a barrel after about 10 days. This reduces the possibility of hydrogen sulphide being produced. When you return the racked wine to the barrel, fill it to within an inch of the lock. And remember, because of evaporation you will need to top it up about once every 3 weeks.

No matter what you've seen in Europe, or what your Italian neighbor does, you cannot run off a bottle of wine every other day from your barrel. When wine comes out, air goes in. Air means oxidation, leading to brown, vinegary, spoiled wine and eventually a contaminated barrel. Perhaps you've heard you can pour olive oil on the top of your wine to delay spoilage, but never do this.

HOW LONG IN THE BARREL?

Taste your wine once a month, and if you think you taste oak, transfer the wine to carboys immediately. Too strong an oak taste will spoil a wine. As long as you don't taste oak, leave the wine in the barrel until you notice a distinct change or softening of the wine — usually after 6–15 months. Don't overdo the barrel aging, it makes for a flabby wine. Wines can stand years in the bottle, but few need more than a year in small (50 gallon) oak barrels.

REDS OR WHITES?

We would not put white wine in a barrel unless we had French or Yugoslavian oak — and then only for 3–6 months. When you put wine into a barrel, you are subjecting it to an oxidative process, and most white wines are reductive wines. Unless you are very experienced, we recommend using barrels for red wines only. And remember, red wine color remains in a barrel permanently; so you can't put a white wine into a red's barrel unless you want a rosé.

STORAGE

Store your full barrel on its side with a lock or waxed bung in the hole. We use a #16 cork in the front of the barrel where the spigot will go, and then sprinkle sulphite around the protruding section. Don't leave the spigot in a barrel; they usually leak. And even if they don't, they offer too much temptation to draw off a glass or two. If you have a cork there, when the time comes to bottle the wine, you can slice the cork off; place the spigot against the cork, and with one blow of a mallet drive the spigot into the barrel, letting the end of the cork float to the top of the wine. It will come out when you wash the barrel.

EMPTY BARRELS

Ideally, you would never have an empty barrel; when you emptied a barrel into bottles, you would be running new wine into it within 48 hours. That way, you would not have to worry about the barrel drying out and falling apart, or going moldy or sour. Unfortunately, it is not always possible for the amateur to have wine ready to go into an empty barrel, and the barrel must be stored; when that happens, you must follow a scrupulous sanitizing procedure:

1. Wash the barrel out immediately. Take it outdoors where you can put the hose in, and flush it until the flushed water runs clear.

2. Return the barrel to its stand and fill it with water. Use 1 tablespoon sulphite crystals and 2 teaspoons of citric acid to every 5 gallons of water. A 50-gallon barrel will require 6 ounces of sulphite and 5 tablespoons of citric acid.

3. Drain and replace the solution every 3 months.

BARREL ALTERNATIVES

If you enjoy a hint of oak in your red wines, but don't feel ready for the responsibility of a barrel, you can use plastic carboys and add oak chips or Sinatin 17 oak extract. Like barrels, plastic carboys let wine breathe, but their walls are so oxygen-permeable that they allow wine to oxidize very easily. This means you must monitor the development of your wine closely. Never leave wine in plastic carboys longer than 3–4 months. Oak chips, too, impart their flavor rapidly; 2–3 months is usually more than enough.

Body for Fruit Wines

One of the disadvantages of fruit wines is the need to add water in order to reduce their acidity and the intensity of their fruit flavor. Additional water results in a wine with little body, a deficiency that may be tolerable in white peach or pear wine, but is definitely unacceptable in red wine of any kind.

There are several ways to compensate for lack of body. You can increase the fruit content until you reach the highest level of acid tolerable (6 g/lit). This would mean using approximately 8 pounds of fruit per gallon, and the result would be a sweet cordial with an intense flavor of fruit. Or you can increase the alcohol level slightly — from 11%, say, to 13%. Or you can use additives, such as dried bananas or glycerine.

But the best solution to the problem is the addition of grape concentrate or chopped raisins to the must. In a 5-gallon batch, as little as 50 ounces (75 ml per liter) of concentrate, or raisins is sufficient.

It is generally agreed that the more complex the flavor and taste of a wine, the better it is. Most fruits contain only 5–6 factors creating flavor and aroma; grapes contain more than 20 and create a much more complex stimulus to the gustatory senses. The addition of grape concentrate or raisins provides the aroma and bouquet essential for a table wine.

Grape concentrate and raisins also offer another advantage. Fruits such as pears, peaches, and blackberries supply very little organic nutrient for yeast growth. Grape concentrate and raisins supply yeast nutrient in abundance. Also, in western Canada and the U.S., blueberries contain a natural yeast inhibitor that will stop fermentation after a level 3–4% alcohol is reached. Add grape concentrate or chopped raisins, and blueberries make an excellent red wine.

Charcoal

You can reduce the color and flavor of a wine by adding activated charcoal. Use activated charcoal as follows:

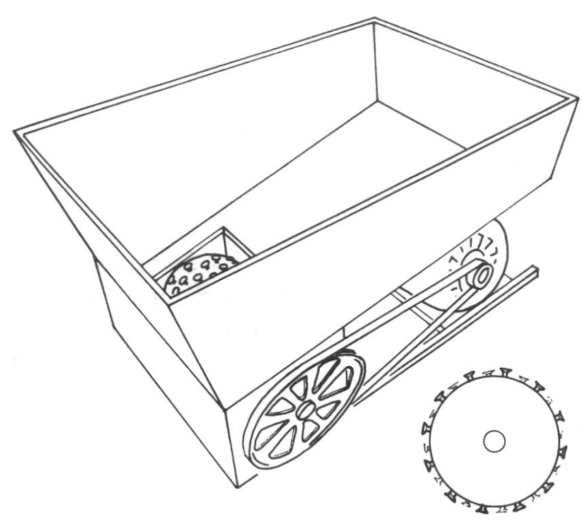

Apple crusher

1. Add ½ ounce for each gallon of wine.
2. Stir the wine twice daily for 3 days, to keep the charcoal in suspension.
3. Allow the charcoal to settle for 6 days, then rack the wine off the charcoal.
4. Add isinglass or bentonite finings and let stand for 10 days.
5. Filter with fine- or sterile-grade filter pads.

Chillproofing

Most wineries chill their wines to 30–32°F (-1–1°C) for a period of 2 weeks. At this low temperature potassium bitartrate crystals precipitate out and drop to the bottom of the container. This prevents any precipitation in the bottle when the consumer puts the wine in the refrigerator. Potassium bitartrate is part of tartaric acid; chillproofing therefore serves to soften high-acid wines, such as those made from hybrid grapes.

For the home winemaker, the coldest part of the garage or the basement in late fall is ideal. If you live in the Sunbelt, a secondhand refrigerator with the shelves removed makes an excellent chillproofing cabinet.

CHILLING WHITE WINES BEFORE SERVING

The best method of chilling white wines is in a wine cooler with a mixture of ice and cold water. Failing that, the refrigerator will do a good job, but never leave white wines of quality in the refrigerator day after day; this impairs the aroma and bouquet, and it can never be restored.

Crushing and Pressing Apples

CRUSHING

To crush apples, you need an apple crusher — a grape crusher will not suffice. There are hand-operated apple crushers, but motorized crushers are more effective. Whether using a hand or motorized crusher, proceed as follows:

1. Clean the crusher thoroughly with a standard sulphite solution before use.
2. Wash the apples. Throw away any rotten fruit.
3. Fill a plastic shaker with half-strength sulphite solution (1 level teaspoon of sulphite

crystals in 1 liter of water) and place it within reach.

4. Place the crusher on top of a plastic tub. (You will need an assistant to hold it in place.) If an alternative is available, do not use your primary fermentor for this.

5. Feed the apples into the crusher 3–4 at a time, taking care to keep your hands free of the crusher teeth.

6. After every 20 pounds of apples, stop and sprinkle approximately ½ ounce of sulphite solution on the crushed fruit to prevent browning. Stir in with a long-handled wooden or plastic spoon.

7. When you have crushed 40 pounds of apples, stop and press them before crushing more.

Apple crushing and pressing are very labor intensive. You will need 20–25 pounds (9–11 kg) of apples to obtain 1 gallon (4 lit) of juice. Apple recipes that do not require crushing and pressing are apple wine (page 60), apple and honey wine (page 58), and scrumpy (page 62).

PRESSING

An apple press is often referred to as a rack-and-cloth press. You place relatively small amounts of crushed fruit in a number of nylon straining bags, pile them one on top of the other, and apply hydraulic pressure. You can adapt a grape press for the job, but it will not be quite as effective.

Operation

1. Clean the press thoroughly with a standard sulphite solution.

2. Most presses will accommodate 5–6 bags, but you will obtain better results with 4.

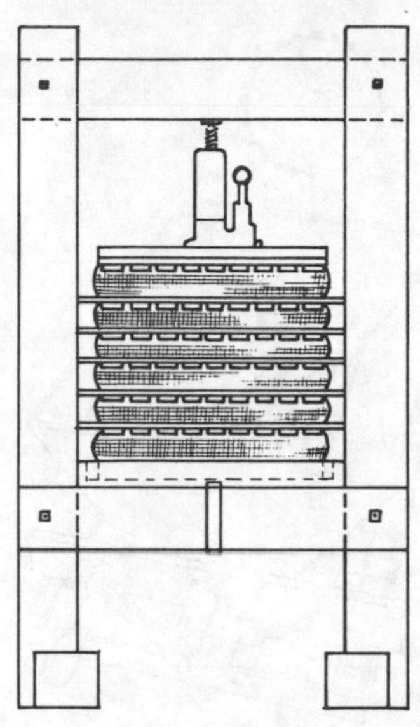

Rack press for apples and pears

Place approximately 10 pounds (4.5 kg) of crushed apples in each bag. Lay the first straining bag in position on the floor of the press, then cover it with the first wooden or stainless-steel pressing plate. Place the next bag in position, then cover it with the second pressing plate. Proceed in this way as if you were making a sandwich: a layer of fruit, then a pressing plate, until all 4 straining bags are in position.

3. Apply pressure to the press slowly.

Crushing and Pressing Grapes

Crushing is designed to break the skin of the grapes and release their free-flowing juice. Pressing is designed to extract what juice

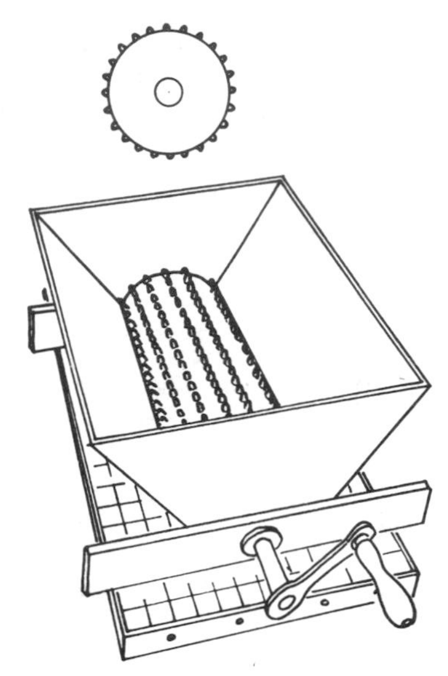

Grape crusher

remains in the crushed grapes by placing them under pressure.

White wines, which are made by fermenting only the juice of grapes, require grapes to be crushed and pressed before fermentation. Red wines, which are made by fermenting crushed grapes in their juice, require grapes to be crushed before fermentation and pressed afterward.

CRUSHING

The most widely used grape crusher consists of a metal or oak hopper feeding onto single or double, manually operated aluminum rollers. You set the crusher on top of the primary fermentor and load the hopper with a lug (35 lb) of grapes at a time. With someone holding the crusher steady, you turn the handle forward then reverse the motion. As the grapes are crushed, they fall through into the fermentor.

A crusher of this design will adequately process up to a ton of grapes — a task of approximately 1½ hours. For larger quantities, an electrically driven model is recommended. Crushers range in price from $90 to $800.

Note: A grape crusher works well with cherries, but is not suitable for apples or pears.

DESTEMMING

Most wineries remove grape stems before fermentation. This reduces the tannin and hastens the aging of the wine. You can remove 80% of your stems with a destemming paddle that you can make yourself. Cut a 4- by 1-inch board into a paddle shape about 4 feet long and hammer six 2-inch nails into the end, as shown. Rake the paddle through the crushed grapes, from bottom to top. The heavy stems, free of grapes, will cling to the nails and can be discarded.

We advise against destemming red grapes entirely. We have experimented with and without stems and consider 80-percent stem removal optimum. Only if your grapes are underripe, and you are worried about a green stemmy taste, should you remove the last 20% by hand.

PRESSING

Traditional grape presses have remained more or less unchanged for a thousand years. Typically, they consist of a threaded steel post mounted in the center of a wooden basket. By turning a steel collar threaded onto the post, the operator can exert a progressive downward pressure. On some models the collar is turned with a simple handle; with others you use a ratchet. Most hand presses come from Italy, but good presses are built in the U.S. also. They come in many sizes, and prices range from $150 to $500.

Presses are rated according to the quantity of crushed grapes the basket will hold. In our opinion, a press that takes less than 100 pounds (45 kg) of crushed grapes at one pressing is a waste of time. If you are processing 100 pounds or less, use a straining bag and press the juice out of the grapes with your hands.

Good presses are made of cast iron, steel, and hardwood, which means they are too large and heavy to load in and out of a car easily. Consequently, presses offered for rent tend to be lightweight and inadequate for the task. While it is true you will probably use a press only once a year, it is easy to justify the $350 investment when the equipment will last a lifetime and can be passed down to your grandchildren.

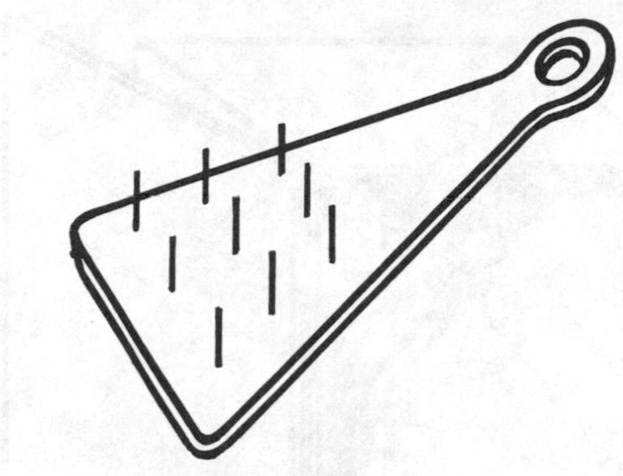

Destemming paddle

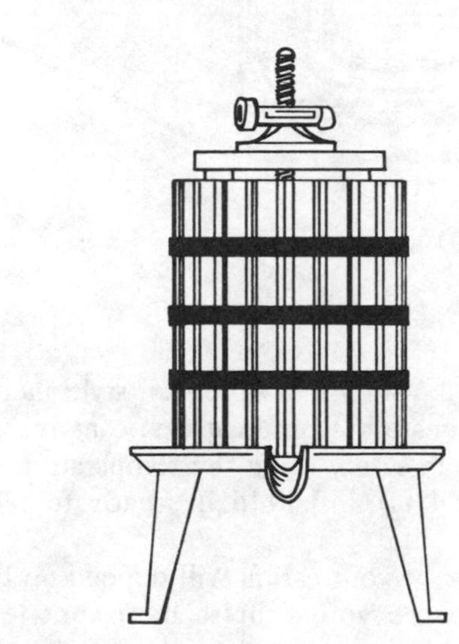

Grape press

OPERATION

1. Clean your press thoroughly with a standard sulphite solution before use. Apply a small amount of Vaseline to the threaded post.
2. Place a container under the spout to catch the juice.
3. Fill the basket to within an inch of the top and place the two halves of the wooden pressing plate on top of the grapes. Place the 4- by 4-inch blocks on top of the plate.
4. Thread the steel collar onto the post and apply pressure to the handle. Perform 2–3 turns and wait while the juice runs. Then perform 2–3 more turns. Repeat until the grapes are reduced to half their original volume.
5. At this point, the turning handle will meet the top of the basket. Turn it counterclockwisc until there is room to insert a second set of blocks on top of the pressing plates. Repeat until 3–4 sets of blocks have been used.
6. When the pomace is condensed to ⅓ its volume, you can consider it pressed. Turn the handle counterclockwise and remove the blocks and pressing plate. Pull out the lock keys from the side of the basket and remove the cake of pomace. You can use this pomace for second-run wine.

Caution: Do not apply extra leverage by extending the handle of the press; you will only break it.

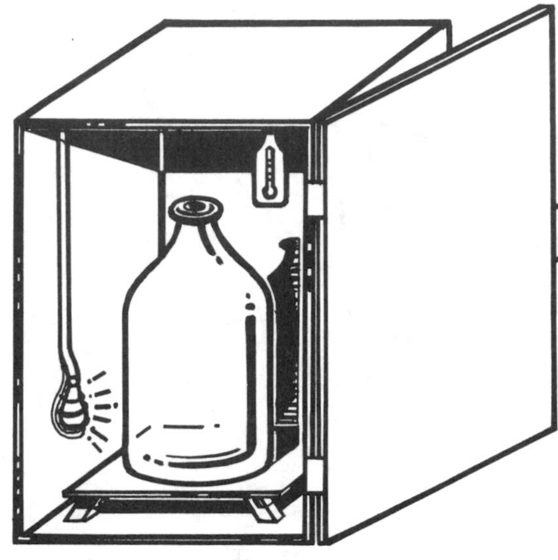

The estufa

The Estufa

For "baking" Madeira or American-style sherry, you need an estufa — an insulated container in which you can safely bring the temperature up to 130°F (54°C) and hold it steady for 2–3 months.

The size of your estufa will depend on the volume of wine you want to bake, but let's assume you have a 5-gallon glass carboy or wooden barrel.

1. Construct a wooden box 39 inches (1 m) square, with a lid or door. Plywood works fine.
2. Line it with fire-resistant insulation.
3. Install a lightbulb or heating pad on the bottom. Take care not to create a fire hazard; if you use a lightbulb, make sure it is enclosed in a wire protective guard, like a mechanic's lamp.
4. To control the temperature of the lightbulb, attach a rheostat to its power source. The heating pad will have its own temperature control.
5. Place a rack over the bulb or heating pad, leaving enough room for air to circulate.
6. Attach a maximum/minimum thermometer halfway up the interior of the box.
7. Place your carboy or barrel on the rack and increase the temperature in the estufa 5°F (3°C) each day until you reach 110–115°F (43–46°C). The slower the baking, the better the Madeira.
8. After 2 months, or when the Madeira has the desired caramelized, baked flavor, bring the temperature down 5°F (3°C) each day until you reach 75°F (23°C).
9. Remove the Madeira.
10. Filter, sulphite, sweeten, fortify, and bottle.
11. Bottle age 1 year.

Fortified Wines

SHERRY

The best-known fortified wine in the English-speaking world is sherry. The name and the wine originate in Spain, at a place called Jerez de la Frontera. The unusual process by which sherry is produced has been carried out for over 200 years. The soil, the method of viticulture, and the naturally occurring yeast by which only true sherry is made are unique to this area of Spain.

Most of the rules for good winemaking are broken by the producers of sherry; air is deliberately allowed access to the wine to promote the natural inoculation of flor yeast. Flor is a film yeast that grows on top of the wine in the presence of air. It protects the wine from oxidation, produces aldehyde, and tends to reduce total acids. It is this flor yeast that creates the typical sherry bouquet and flavor.

Another unique factor is the solera method, which permits the sherry maker to market an identical wine year after year. Briefly, a solera is a series of butts, or barrels of 130-gallon capacity, grouped in 5 stages, or levels. They are not necessarily set vertically, but it helps to think of them that way. Picture 25 barrels on 5 shelves, 5 on each shelf. The 5 at the top contain the newest wine, and the 5 on the bottom contain the mature wine, of which 50% is drawn off to be blended and bottled. As the wine is drawn off from the bottom, the same amount is moved down through each level of the solera and new wine goes in at the top. No barrel is ever emptied, so what is withdrawn from the bottom is a mixture of every wine that ever went into the barrel, most of it 5–10 years old. As you can imagine, this system is not practical for the home winemaker and too labor-intensive for most American wineries.

Fino, Amontillado, Oloroso

Even in Spain, the development of flor yeast on top of the wine is not certain. The wine with a successful flor is called fino; it is light in color and usually sold as a dry aperitif. The next classification has a flor but will not quite reach the same standard; it's called amontillado. It is sold dry or slightly sweetened. Many people prefer this over a fino.

The third class, oloroso, is quite different. Dark and heavy, with a great aroma, it is more heavily fortified and sweetened. Oloroso is marketed under the names cream sherry, shooting sherry, or hunting sherry and is much favored by the English squires.

The grape used to make sherry is palomino; the grape used to sweeten it is Pedro Ximenez. The wine is plastered, which refers to the addition of gypsum to reduce the pH. The flor prefers a temperature of 65°F (19°C) and develops in a wine with 14.5–15.5% alcohol by volume, a pH of 3.2–3.5, and a titratable acid of 4–5 g/lit. After the desired sherry flavor is attained, the wine is blended and fortified with a strong, aged brandy. Fino and amontillado are fortified to about 17.5% alcohol by volume; cream and oloroso are fortified to 20%.

We are reluctant to encourage anyone to make flor sherry; there are many factors to control, and because you must deliberately aerate your base wine, you stand the chance of losing it to spoilage bacteria or oxidation if the flor does not develop. We have had some successes in the lab, but not consistently. And we've seen a great deal of spoiled wine that people thought was flor sherry. We have even met a professional enologist who insisted that sherry flor was merely mycoderma — little wonder he didn't like sherry!

California Sherry

Some excellent sherry is produced outside of Spain. Flor sherry is made in California, Canada, and New York state, as well as in Australia and South Africa. But we much prefer to recommend the production of American-style sherry; it's easier to make, with little or no chance of failure.

American sherry gets its sherry-like flavor from being "baked," or heated for several months. This process is said to have originated when Portugal was shipping white wine to India in barrels that sat in the tropical sun for weeks on board sailing ships. The resulting product was so much like sherry that when the British fell out with the Spanish, they turned to their Portuguese friends to supply the next best thing. It was called Madeira after the island from which it came, and the Portuguese soon learned to copy the conditions of the tropical boat trip for the British market.

They built special aging cellars where the fermented wine could be heated to 130°F (54°C) for several months, until it acquired the baked, caramelized flavor that was so like sherry. The special baking cellar is called an *estufa* (or "stove") and is quite simple to create in your own home. (See "The Estufa," page 252.) Madeira is produced as dry, medium, or sweet and is also fortified. Some Madeira has been known to keep for more than 75 years.

Most California sherry is produced in this manner, and much of it is a very acceptable substitute for Spanish sherry, which is now very expensive. You can try for a flor sherry, but if the flor does not develop within 2–3 weeks after fermentation, you must get your base wine fermenting again to increase the alcohol to 17%, then sweeten it slightly. You need at least 2% residual sugar to develop the caramelized flavor. When a wine of 2% sugar is fortified to 17% alcohol, it tastes very dry as an aperitif, so don't worry about having too sweet a wine. In fact, we suggest sweetening to about 4% for the average palate.

Any sound, low-acid white wine except Muscat will do as a base for American-style sherry production.

Vermouth

Vermouth is a flavored, fortified wine, drunk usually as an aperitif. It is a social wine, best served on the rocks with a twist of lemon or a cocktail additive. Traditionally, French vermouth is white and dry, and Italian vermouth is red, tawny, and sweet. Martinis are made with dry vermouth; Manhattans are made with sweet. When we were younger, our favorite was a "triple" — one part French vermouth, one part Italian vermouth, and one part gin.

More than 80 herbs, spices, and natural flavors are used in the production of vermouth. Berries, barks, seeds, roots, fruit peel, leaves, plants, flowers — everything from angelica to wormwood.

Wormwood was the principal ingredient in absinthe, the famous aperitif that was said to destroy the minds of famous writers and poets in Paris during the twenties. Vermouth won't destroy your mind, but I suspect it of being a mild aphrodisiac, especially Italian vermouth.

Vermouth is easy to make. It requires only 17–18% alcohol by volume, and you can usually ferment to that level. Vermouth herbs are available from winemaking supply stores already blended and with instructions for use. Basically, you make a sound, high-alcohol wine from a neutral grape concentrate, then immerse the herbs. You will find a vermouth recipe on page 114.

Port Wine

Port is named after the city of Oporto, where it was shipped from Portugal. It is a fortified dessert wine that can be white, red, or tawny, with varying degrees of sweetness. A dry white will have 4% residual sugar; a ruby port may have as much as 14% residual sugar. The sweetness is not oppressive, because the high alcohol level offsets it. Port is fortified to 17–20% alcohol by volume. Most Portuguese port (the only true port) has 20% alcohol by volume.

Port is produced by fermenting grapes to 6–8% alcohol, then stopping the ferment with the addition of high-proof brandy. This brings the alcohol content up to the desired level and leaves the desired sweetness. The wine is then aged in large oak barrels until an evaluation is made by the cellar master. If it is unexceptional, it will be bottled after 2–3 years and sold as ruby port. If it shows promise, it will be aged longer — as much as 15 years — and sold as vintage port. Vintage port is usually tawny in color, with a sweetness that is inexplicably diminished.

Port is one of the world's great wines. Lately, however, it has fallen out of fashion as we have become preoccupied with avoiding calories. In North America, it is served mainly during the Christmas season as a complement to shortbread and cheese for dessert. You will find a recipe for port on page 168.

Fortifying Wines with Distilled Alcohol

You can bring your sherry, Madeira, port, and vermouth wines up to the desired 17–20% alcohol strength by syrup feeding, or you can fortify them with distilled alcohol; that is, brandy, vodka, or whiskey.

Commercial wineries generally choose to add alcohol. In Portugal, where port is produced, they ferment to 6–8% alcohol, then add strong brandy, 65% alcohol by volume, to halt the fermentation. This results in 4–14% residual sugar and 20% alcohol by volume.

For the home winemaker, however, it is more economical to ferment to 16–18% alcohol by volume to avoid the expense of using large quantities of brandy to fortify the wine. Generally speaking, the drier the fortified wine, the lower the alcohol content needs to be. Fino, or dry, sherries contain 17% alcohol by volume; cream, or very sweet, sherries, on the other hand, contain 20% alcohol by volume. Use the Pearson Square, below, to determine the amount of alcohol you should add to your wine.

When you fortify a wine, buy the highest-proof alcohol available. Your aim is to strengthen the wine without diluting the flavor. The regular strength of distilled liquor is 80 proof, or 40% by volume (U.S. measure). But in Canada you can usually buy 100 proof, and in the U.S. 130 proof is generally available.

THE PEARSON SQUARE

The Pearson Square helps you calculate the amount of spirit you need to fortify a wine.

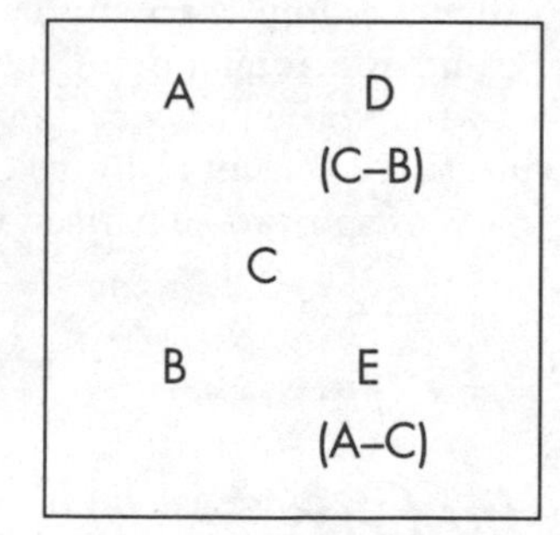

A is the alcohol content of the spirit you intend to use.

B is the alcohol content of the wine you intend to fortify.

C is the alcohol content you intend your fortified wine to have.

D equals C minus B. The difference is the proportion of fortifying spirit you will need.

E equals A minus C. The difference is the proportion of unfortified wine you will need.

Growing grapes in your backyard

If, for example, the fortifying spirit has an alcohol content of 40%, your wine has an alcohol content of 15%, and you want to fortify your wine to 20%, you simply subtract B (15) from C (20). The remainder D (5) is the proportion of spirit you will need. You then subtract C (20) from A (40), and the remainder E (20) is the proportion of wine you will need. Thus, to fortify your wine to 20% alcohol by volume, you will need 5 parts of spirit to 20 parts of wine.

Growing Grapes in Your Backyard

Our personal vineyard is feasible and affordable. Thanks to the dedicated research of the commercial growers, much of the uncertainty has been removed from amateur viticulture. There is now a variety of hybrid vines to choose from and a wealth of information about their cultivation and grape yields. By consulting the tables, you can select vines to suit the needs of almost any geographical location. The American Wine Society and your local state/provincial department of agriculture can provide additional information on growing grapes under difficult conditions.*

Ten vines will provide you with approximately 150–200 pounds of grapes each year, depending on the variety. They will start to yield within 3 years and will reach full productivity after 5 years. At the end of 30 years, their quality and yield will diminish.

The ideal location is a solar greenhouse, but a plastic tent constructed of 6-millimeter polyethylene, or any of the large plastic-fiber greenhouses sold at nurseries for home assembly, will provide heat for your vines and protect them from late-spring and early-fall frosts.

Black plastic ground cover or mulch helps keep the soil warm and keeps the weeds and grass down. As soon as the ground under your tent or greenhouse warms up to 50°F (10°C), your vines will start to bud, but a severe late-spring frost could cost you your crop. A string of low-wattage lightbulbs kept on during the night will protect your young budding vines and give you a longer growing season. A maximum/minimum thermometer and a light timer switch will give you this needed protection in spring and fall.

WATERING AND FERTILIZATION

Many types of self-watering systems are on the market. Our best results have been obtained using 45-gallon plastic drums that gravity-feed a buried weeper hose through a spigot at the bottom. The spigot allows you to control the water flow, and you simply refill the drums with a garden hose every few days. For a 20- by 20-foot greenhouse, two drums are quite sufficient.

*See "Buyer's Guide," page 281.

Before adding fertilizer, have a soil test done. If water-soluble fertilizer is available, it can be added to the watering drums. In British Columbia, where our home vineyard is located, ½ pound of Sulpo Mag and ⅙ pound of 11.55.0 applied to each plant in mid-April and mid-May has worked very well.

PESTS AND DISEASE

Vines are vulnerable to pests and disease. Regions will differ, but in British Columbia cutworms and leafhoppers pose the greatest insect threat. Vines also need protcction from powdery mildew and bunch rot.

Birds arc not a problem for the greenhouse vintner, but grapes outdoors are at risk — Maréchal Foch seems to be especially susceptible. Netting your vines in September and October will prevent serious loss.

VENTILATION

Special attention must be given to airflow and ventilation. A thermostatically activated fan to pull hot, moist air out of the greenhouse is very beneficial.

PLANTING

New vines should be planted with the graft 1–2 inches above ground level. The roots must be trimmed to 6 inches in length and placed in the ground with wet peat moss and a handful of blood and bone fertilizer.

TRAINING AND TRELLISING

The vine is unruly; unless trained over a wire trellis and pruned systematically, it will dissipate its growth in a luxuriant tangle of

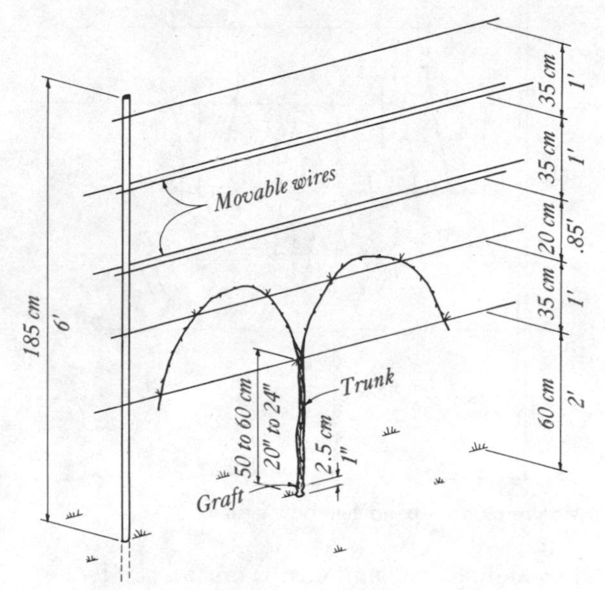

Before growth in spring

shoots that yield grapes of little sugar and poor taste. The object of training and trellising is to encourage a moderate, disciplined growth that will produce grapes of good flavor. As an amateur, you have the luxury of ignoring quantity to concentrate on quality.

A diagram of the pendulum bow system appears on next page. The permanent trellis is 6 feet (1.8 m) high and provides a framework of 7 wires over which the vine is trained. Three are fixed; 4 can be raised to accommodate the growth of the plant. Your job for the first 3 years is to restrict growth by selective pruning and to direct the most promising shoots upward and into the trellis.

First Year

The first year, the goal is upward growth. Each plant must be directed to grow as tall and straight as possible. Prune off all side shoots and tie the stem to its individual support stake.

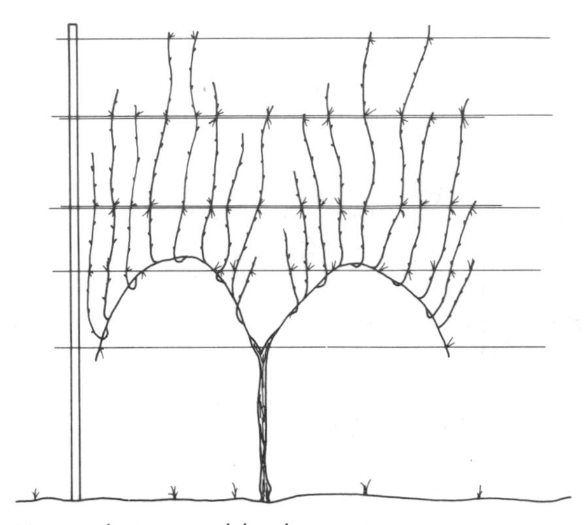

Mature plant — pendulum bow system

Depending on conditions, as much as 10 feet of growth can be expected in the first season.

Second Year

In about February, when no more hard frosts are expected, prune the stem back to the height of the first wire. A number of buds will have appeared along the stem. Select 3–5 buds and prune off the rest. What you are doing is conserving and directing the growth of the vine. If a plant appears weak and has few buds, leave only 3; if a plant appears vigorous and there are many buds, select 5. As they develop, encourage the new shoots to grow as vertically as possible by tucking them into the wires of the trellis.

Third Year

Prune the main stem back to the height of the bottom wire. The main stem must be kept at this level; particularly vigorous plants need to be topped again in late August. Of the shoots you selected to remain the year before, choose 1–2, depending on the vigor of the plant, and prune off the remainder. Tie the arms over the second wire, then bend them back and attach their tips to the first wire in bow fashion, as shown. These arms will now have buds of their own. Select 10–15 on each arm, depending on the vigor of the plant, and remove the rest. As they grow, tuck them into the wires of the trellis to encourage vertical growth.

Fourth and Succeeding Years

When no more hard frosts are expected, prune the main stem back to the height of the first wire and keep it free of all shoots below the main arms. Keep the arms tied in a bow shape. With the vine now established, these arms can support more shoots. Select additional buds to make a total of 18–20 shoots and buds on each arm, including those from the year before. Prune off the rest.

As the new shoots grow, continue to tuck them into the trellis wires and tie them in a vertical position. By now many shoots will be thrusting upward above the trellis. Keep them to a height that will allow them to bear grapes without breaking off. Top them as necessary. In June, remove leaves from around hidden grape clusters to prevent mildew.

Winter Cover

Where winter temperatures fall below –4°F (–18°C), your vines will need winter protection. Six inches of earth, a dry mulch, and a good snow cover should keep them safe, even down to –20°F (–20°C). The younger the vines, the more susceptible they are to frost damage. If you live in a severe winter region, therefore, you may need to keep the trunks shorter — perhaps as low as 12 inches — so that the total height will be only 4 feet. This will allow you to cover the trunks entirely during hard frost.

CHOOSING A HYBRID

If you live in a marginal grape-growing area, you will need to select from the new range of hardy North American hybrids. The information that follows was obtained from the Agriculture Canada research station in Summerland, British Columbia. Canadian growers face enormous obstacles: a climate unsuitable for grapes, plus the highest production costs in the world. Although none of their hybrids can yet attain the sugar and acid balance necessary to produce a drinkable wine without amelioration, Canadian growers have made tremendous advances in their search for the perfect vinifera clone.

Characteristics of Some Grape Varieties

Variety	*Species (1)*	*Ripe date (2)*	*Color (3)*	*Fresh fruit (4)*	*Use (5)*	*Hardiness (6)*	*Production (7)*	*Powdery mildew (8)*
Aurora (S-5279)	H	Early	W	F	W	P	5	M
Baco Noir	H	Early	B	P	W	G	10	M
Cayuga White	H	Late	W	F	W	G	10	S
Chelois	H	Late	B	P	W	F	7	M
Diamond	L	Midseason	W	F-G	W	G	5	M
De Chaunac (S-9549)	H	Late	B	P	W	G	10	M
Festivee (V-53033)	H	Late	B	G	F	F	10	M
Foch	H	Midseason	B	P	W	G	10	M
Gewürztraminer	V	Late	W	P	W	G	5	Se
Himrod	H	Early	W	G	W-T	P	4	Se
Okanagan Reisling	?	Late	W	G-F	W-T	F	6	M-Se
Rougeon (S-5898)	H	Late	B	P	W	G	7	M
Rulander	V	Late	W	P	W	F	7	Se
Schwartz Riesling	V	Midseason	B	P	W	F	7	M
Sovereign Coronation (#361)	H	Early	R	G	F	G	10	S
#166 (Sov. Opal)	H	Late	W	P	W	G	6	S
Valerien (SV. 23410)	H	Late	W	P	W	G	10	S
Verdelet (S-9110)	H	Late	W	G	W-T	P	8	M-Se
Veeblanc (V-53263)	H	Late	W	P	W	G	10	S
V-49404	H	Late	W	P	W	F	5	Se
V-50201	H	Late	W	P	W	F	10	M
Weissburgunder	V	Late	W	P	W	F	10	M

(1) L = labrusca; V = vinifera; H = hybrid
(2) Early: mid-August – mid-September
Midseason: mid-September – first two weeks of October
Late: second week of October onward
(3) W = white; B = blue; R = red
(4) G = good; F = fair; P = poor
(5) W = wine; T = table
(6) G = good; F = fair; P = poor
(7) 1 = poor; 10 = best
(8) S = slight; M = moderate; Se = severe

Agriculture Canada's *Grape Production Guide '85* states, "Varieties such as Concord, Bath, Delaware, Campbell Early, de Chaumac, Foch, Patricia, Sheridan, Buffalo, Portland, Niagara, Elvira, Fredonia, Veeport, and Veeblanc V-53263 and V-50201 are hardy, but planting these varieties for wine use is no longer recommended."

RECENT EXPERIMENTAL PLANTINGS

Encouraged by a similarity between the climates of marginal grape-growing areas in North America and areas of Germany where some of the best white wines originate, a project headed by Professor Helmut Becker of the Department of Viticulture in Geisenheim, West Germany, carried out experimental plantings in 1977–1985 in Kelowna and Oliver, the major grape-growing areas of British Columbia.

Forty-two varieties were chosen, including 6 vigorous wine hybrids developed at the Summerland research station in the same area. Scionwood, rootstocks, and grafted plants were imported for the experiment, and 33 Geisenheim selections were grafted to one or more of the rootstocks supplied, making a total of 74 different combinations for consideration.

The two sites chosen had a western exposure and were 62 miles (100 km) apart. Their climate averages are listed below.

Last day of spring frost		*First day of fall frost*		*Frost-free days*	*Temperature*		*Total heat units above 10°C*
					Min	*Max*	
Site #1				192	−4°F	97°F	
Apr 17	30°F −1.0°C	Oct 31	30°F −1.0°C		−19.6°C	36.1°C	1,179
Site #2				197	0°F	97°F	
Apr 52	9.5°F −1.4°C	Oct 21	28°F −2.0°C		−18.3°C	36.2°C	1,298

A list of the most promising cultivars and clones chosen on a combined basis from this ongoing experiment follows in detail.

Gewürztraminer 15GM
Auxerrois KL22
Osteiner
Geisenheim 4-46
Schonburger
Weissburgunder 5GM
Ortega
Ehrenfelser
Geisenheim 32258
Müller-Thurgau KL Heinz

The wines were assessed, and their pH, phenolics, and glucose/fructose sugar and malic/tartaric acid ratios were analyzed. All were given a minimum of three tastings without their identity being known to the panelists and were scored according to the Modified Davis System with a 20-point maximum score. The most promising were

Auxerrois	Ehrenfelser
Müller-Thurgau*	Schonburger
Ortega	Gewürztraminer
Weissburgunder	Scheurebe

*Müller-Thurgau clone from Davis proved to be superior to the MT in the project.

Summary of Wine Scores

Wine	Score average	Comments
AUXERROIS	13.2	Faint fruity nose; low acidity; fair body
SCHONBURGER	12.0	Clean, flowery, fruity; light muscat nose, slightly acidic; light muscat flavor; stemmy flavor; fair body
WEISSBURGUNDER	13.4	Clean, light, neutral, pleasant; balanced; light fruity flavor
MÜLLER-THURGAU (11)	11.9	Slightly musty, fruity nose; thin, light muscat flavor with an unpleasant note
MÜLLER-THURGAU (1/1)	11.5	Clean, faint, fruity nose; bland, thin flavor; lacks body
EHRENFELSER	11.6	Light, fruity, muscat nose, with a spicy note; slightly sulphury; sharp acidity; full-bodied; strong, fruity flavor
ORTEGA	11.7	Clean, light, fruity; muscat nose; balanced acidity; unique fruity, soft, light muscat flavor
GEWÜRZTRAMINER	12.1	Fruity, slightly sulphury nose; acidic, slightly astringent; fruity flavor with a spicy note; acceptable body; slightly sweet
SCHEUREBE	11.3	Fruity, ripe; unique vinifera nose; slightly acidic; slightly astringent; attractive flavor; slightly sweet

In March 1984, propagation of six of the most promising cultivars began, namely:

Schonburger
Ehrenfelser
Ortega
Weissburgunder 5GM
Müller-Thurgau*
Auxerrois KL56*

Cuttings may not be available in all areas. Government departments of agriculture, horticulture, or viticulture are good sources of information, or write to the American Wine Society for up-to-date information on purchasing grapevine cuttings for your climatic region.†

*These were not in the project, but were superior clones.

†See " Buyer's Guide," page 281.

Parentage and Brief Description of Cultivars Tested

AUXERROIS: Named in honor of the earldom of Auxerrois in northern Burgundy; a clone of Pinot blanc selected for its improved sugar-to-acid ratio at maturity; strong growth, light yellow berries, very good wood maturity

SCHONBURGER: Spätburgunder X 1P1 (Gutedel X Muscat Hamburg) Italian table grape by Pirovano; thick-skinned berries, traminery type; strong aroma, average-to-vigorous growth; fair wood maturity

WEISSBURGUNDER: Clone of Pinot blanc; thin-skinned berries, neutral flavor; moderately strong growth

Parentage and Brief Description of Cultivars Tested

MÜLLER-THURGAU: A vinifera cross from the Thurgau area of Switzerland, selected in 1882 at Geisenheim by Professor H. Müller; now believed to be a Riesling X Sylvaner or a Riesling X muskateler cross; large berries, light muscat aroma, strong growth; poor wood maturity

EHRENFELSER: Riesling X Sylvaner Geisenheim 1929; Riesling type, average-to-strong growth; very good wood maturity

ORTEGA: Müller-Thurgau X Siegerrebe, Wurzburg, 1971; vigorous growth, many leaves, moderately large berries, muscat aroma; poor-to-fair wood maturity

GEWÜRZTRAMINER: Probably from Tramin in South Tyrol; small berries, unique varietal bouquet, moderately strong growth; very good wood maturity

SCHEUREBE: Sylvaner X Riesling Alzey 1916; very strong growth, moderately large berries, pronounced bouquet; fair-to-good wood maturity

Note: Although Maréchal Foch was not part of the above project, it is one of the most popular grapes grown by winemakers and its parentage may be of interest.

MARÉCHAL FOCH: Kuhlman 188-2, derived from Kuhlman 101-14 X Gold/Riesling; small, deeply colored berries; very productive, vigorous; the most widely planted blue grapes in B.C. and Canada; very susceptible to bird damage; excellent wood maturity

Hybrid Grape Use

The harsh climatic conditions of Canada and the eastern U.S. oblige growers in these regions to resort to hybrid vines that produce grapes high in acid and low in sugar. Essentially, they are a compromise; they make a wine that is not as good as good wine from *Vitis vinifera*, but better than a poor wine from vinifera. A cruel critic might say they are "better than no wine."

Ideally, a red table wine should have 22% sugar and 5.5–6.5 g/lit acid, while a white table wine should have 21% sugar and 6.0–8.0 g/lit acid. In a good year hybrid grapes will have 17% sugar and 12 g/lit acid. Obviously, it makes more sense to use hybrid grapes for white wine. But even for a white wine, most hybrid grape musts will still need amelioration. You can ameliorate in three ways: You can dilute the acid with water and add more sugar; you can neutralize the acid with chalk and add more sugar; or you can add grape concentrate, which is itself naturally low in acid and high in sugar.

In our experience, hybrid grape growers have already added too much water as irrigation to get their 5–8 tons per acre. More water only makes thinner wine — in a must already too thin. Nevertheless, we offer a recipe for hybrid grapes ameliorated with sugar and water on pages 204–206.

Chemically neutralizing high acid is a better option, and we offer a recipe for hybrid grapes ameliorated with sugar and chalk on page 196. If you want to make your own recipe using chalk, see "Lowering the Acid Level," on page 222.

But the best amelioration for a high-acid, low-sugar hybrid grape must is the addition of grape concentrate. Grape concentrate loses most of its acid during the process of concentration. A typical concentrate will have an acid reading of 2.5–3.0 g/lit at a specific gravity of 85 (1.085). If your grape must has an acid content of 12 g/lit, adding an equal quantity of concentrate will reduce it to 7.3 g/lit. However, you must buy concentrate in which the acid level has not been adjusted; producers often make up the acid level of their concentrates with citric acid. Always read the container label carefully.

Most grape concentrate has an SG of 1.337 (68 Brix). Forty-two ounces (1.3 lit) added to 5 gallons (20 lit) of must will increase the SG to 1.020, or 5 Balling. We provide a recipe for hybrid grapes ameliorated by grape concentrate on pages 198–200.

If the acid content of your grapes is not *too* high — say, 8 g/lit or less — you might want to increase the sugar and body by adding grape concentrate instead of sugar.

Because they are underripe, hybrids have more pectin and more malic acid than mature vinifera grapes. To obtain a better release of juice at crushing and a greater clarity after fermentation, add 1 teaspoon of pectic enzyme per 100 pounds of grapes as you crush, and inoculate the secondary fermentor with Leucostoc to reduce the malic acid with a malolactic fermentation. Remember, too, that sulphite acts more effectively on underripe grapes; so use only 3 teaspoons in 3 cups of water for 12 lugs of grapes.

Most white wines made from hybrid grapes will also be improved by the addition of 2–3 ounces per gallon of wine conditioner before bottling.

If you are making red wine from hybrids, amelioration is even more important. And take care to remove all the stems from the must. When grapes are underripe, the stems are green and impart a "stemmy" taste to wine.

Chillproof all hybrid grape wines.

Light Wines

Light wines are new to North American wineries, but the Germans have always made wines with 7–9% alcohol. Until recently, California law required a minimum of 10% alcohol in table wine, but the demand for less calories as well as less alcohol finally brought a change in the law. While light wines do sell, they are not as successful as the wineries anticipated. Regular-strength wines are as easy to make as light wines; so why not just add water or soda to your wines and make spritzers for your friends who want less calories?

Purchasing Fresh Grapes

SOURCES

While many growers throughout Canada and the USA may contradict us, we will say that if you can get fresh wine grapes from northern California, then do so. That is where they consistently harvest premium varietal grapes, perfectly balanced. California grapes are normally available September 5–21 (Cucamonga, near Los Angeles) and October 1–15 (Napa Valley). Grapes from the warmest areas of California are not good for table wines. U. C., Davis, lists 5 California growing areas by temperature. Napa and Sonoma are best; Bakersfield and the Los Angeles area are poorest.

If there is a winemaking supply store in your area, it should be able to tell you where to find a reputable importer. Otherwise, contact your local chapter of the American Wine Society or a local winemakers' club.

You can also check with the wholesale produce importers listed in the *Yellow Pages*. They are less reliable because they will be bringing in Central Valley grapes, but they will promise you anything you ask for. Unfortunately, you won't be able to tell a Zinfandel grape from a Cabernet just by looking at it.

Don't discount Central Valley grapes from Lodi, California. Zinfandels, Chenin blanc, and so on, are often nicely balanced from that area, and much cheaper than Napa Valley grapes. If you like Zinfandel, or blanc de noir from Zins, you will be happy with Lodi grapes.

Although we consider wine grapes grown in Canada and the northern USA suitable only for white wines, there are microclimates, especially in Oregon, which can give you exceptional grapes in a good season. There are also considerable new plantings in Arizona, New Mexico, and Texas. These plantings are so recent that there is little information about them. If you live in one of these states, contact your state department of agriculture to find out what varieties are available. The climate of these states produces enough heat units for good red wine grapes.

SHIPPING CHOICES

Fresh grapes and juice are now shipped in a variety of packaging. Traditionally, California grapes have arrived in wooden crates, or lugs, that hold 33–36 pounds (15–16 kg). But you can now buy crushed and destemmed grapes in 5-gallon pails and 50-gallon drums. These are held and shipped under refrigeration from any grape-growing area in North America, including the Napa Valley. They are available from September 10 to November 15. There are potential problems with this form of packaging, however, because if the storage temperature is not low enough, the grapes start fermenting before they are sold. You end up with half-fermented grapes or juice, filled with wild yeasts and bacteria.

The third alternative is the relatively new aseptically packaged juice, which retains the fresh fruit flavor and aroma. It has no yeast or bacteria growth and does not need to be refrigerated. This is the best deal of all for those who want to make less than 20 gallons, or who have limited space for winemaking. It may be the best way of making white wine, because you do not have to spend time processing the grapes, which tends to oxidize them and change the color and flavor of the finished wine.

Grape juice from France, Italy, and Germany is also imported to eastern Canada in refrigerated tanks. The importers claim to be selling varietal juice, but of course, this is very hard to confirm unless you have years of experience with the aroma of fresh grape juice. This method of importing also presents a hazard: Unless the temperature of the unpasteurized juice is rigorously controlled during shipping and packaging, fermentation can occur.

Problems and Solutions

PROBLEMS, CAUSES, AND SOLUTIONS — QUICK REFERENCE

Problem	Cause	Solution
Bitter taste.	• Wine oxidized: left too long in primary or too much air in secondary. • Acid too low.	• Prevention only (p. 269).
Ferment won't start. (Check with hydrometer first — ferment may already have finished.)	• Not warm enough. • Poor wine yeast. • Too much sulphite. • Too much acid.	• Apply heating pad (p. 269). • Add more yeast (p. 269). • Reduce sulphite (p. 238). • Reduce acid (p. 222).
Ferment won't stop. (Stuck ferment.)	• Must too cold. • Too little nutrient.	• Apply heating pad (p. 269). • Add ¼ tsp nutrient and stir.
Foaming.	• Low alcohol.	• Use antifoaming agent.
Cloudiness in bottle.	• Bacterial contamination.	• If wine still smells and tastes OK, sterile-filter and sulphite.
Liquid yeast starter won't start.	• Outdated yeast. • Too cold.	• Add pkt dry wine yeast. • Warm to 80°F (27°C).
Mycoderma.	• Contamination of airspace in carboy.	• Sterile-filter, sulphite to 50 ppm, and top up (p. 269).
Not enough wine for topping up.	• Bad recipe. • Too much lees.	• If fermentation has ceased, add a similar wine (commercial if necessary). • If amount needed is less than 4 oz per gallon (30 ml per liter) add cold tap water. • Add sugar and water. • Move to smaller containers.

PROBLEMS, CAUSES, AND SOLUTIONS — QUICK REFERENCE (Cont.)

Problem	Cause	Solution
Ropiness (long strands visible in wine).	• Bacterial contamination.	• If wine still tastes and smells OK, agitate to break up bacteria chains, sterile-filter, and sulphite.
Rotten-egg smell.	• Too fast a ferment. • Late racking. • Sulphur spray on grapes.	• Avoid Montrachet yeast. • Aerate (splash vigorously) and add 50 ppm SO_2 as many as 3 times (p. 267).
Sediment in bottle.	• Sucking up lees when racking. • Failture to add sulphite when bottling. • Tartaric acid crystals.	• If soft sediment: Fine, filter, and sulphite. • If hard crystals: Just decant before serving.
Acid level too high.	• Underripe grapes. • Measuring error.	• Reduce acid (p. 222).
Unstable in bottle.	• Bottled before fermentation complete. • Bulk aging omitted.	• Fine, filter, and sulphite; add sorbate (p. 268).
Won't clear.	• Not enough tannin. • Bacterial contamination. • Lack of pectinase.	• Add ½ tsp liquid tannin per gallon (4 lit), fine with bentonite or sterile-filter, and sulphite. • Add pectic enzyme and warm to 80°F (27°C) (p. 22).
Won't sparkle.	• Alcohol too high. • Not enough yeast in suspension. • Too cool.	• Use Roto-Gas method (p. 214). • Add fresh *tirage*. • Move to warm place and agitate daily.

ACETOBACTER

The Problem

You don't need flies to get acetobacter in your wine. Vinegar bacteria is as omnipresent as wine yeast. There is almost always some acetic acid in wine; in fact, in minute quantities, it improves the bouquet and the complexity of wine. However, like salt, a little is great; a lot is a disaster. The amateur will get enough vinegar without trying; so the object is to make sure it is not enough to smell or taste like vinegar.

If you wonder if that unpleasant smell in your wine is vinegar, go to the kitchen and take a good sniff of household vinegar, then go back and check your wine. If you still think you have vinegar in your wine, you probably have.

You can either throw it all away — including the barrel — or, if you use a lot of wine vinegar, you can take a chance that the acetobacter in your wine is a good strain and make 40 gallons of wine vinegar. First move it from your winemaking area. Add 40% water and make sure it gets lots of air but not insects. Cover the container with a mesh and stir it regularly. In 2–3 months, depending on the temperature, fine and filter the vinegar and bottle it like wine.

Prevention

There is no solution to acetobacter. Prevent it by using cultured wine yeast, maintaining 30–50 PPM SO_2 at all times, and never stopping fermentation before 10% alcohol is attained.

HYDROGEN SULPHIDE

The Problem

Hydrogen sulphide (H_2S) should not be confused with sulphur dioxide (SO_2). Hydrogen sulphide smells like rotten eggs or a tropical lagoon when the tide is out. You often smell it when you're downwind from a pulp mill. Hydrogen sulphide in wine is a very common problem, even among commercial winemakers. Just last year, we bought a premium California wine containing hydrogen sulphide. How it escaped quality control is puzzling, because hydrogen sulphide is not created in the bottle. What is most surprising is the fact that so little reference to the problem can be found in existing literature on winemaking.

Hydrogen sulphide is created by wine yeast during fermentation, and some strains, such as Montrachet, produce more H_2S than others. Generally, the problem is not apparent until the secondary fermentation stage. During the secondary fermentation, dead yeast cells build up in the sediment, and as they break down, their natural sulphur content is reduced to hydrogen sulphide. The heavier the deposit, the warmer the temperature — and the greater the probability of hydrogen sulphide developing. If hydrogen sulphide remains in wine more than a week or so, it can be converted to mercaptans, which smell even worse than H_2S; the odor is then referred to as skunky or garlic-like. But that's not the end of it: If left long enough, the mercaptans become disulphides, which smell even worse.

The Solution

The texts on commercial winemaking all claim that the removal of H_2S is easy with copper sulphate or a commercial product called Sulfex O. The company that distributes Sulfex is reluctant to release it for sale to home winemakers, so it cannot be rated by us; however, the success rate of copper sulphate leaves much to be desired.

Most enologists claim that if you catch it early, you can get rid of hydrogen sulphide by aeration (splashing the wine from one container to another) and adding 50 ppm of sulphur dioxide. We suggest doing this at least 3 times at weekly intervals.

We have never tasted a good wine that has been cured of severe hydrogen sulphide during the secondary fermentation. There is always a hint of sulphide in the nose and a bitter finish on the palate.

Unfortunately, the only real answer to the problem is to do everything to *prevent* H_2S. Avoid yeasts that produce excess hydrogen sulphide. After the initial racking from the primary into your carboy, do not leave your wine on the sediment more than 10 days; syphon off in less time, if possible. If the wine is still fermenting, do not rinse the carboy with sulphite solution, just make sure it is topped up with tap water.

INSTABILITY

The Problem

A wine is said to be unstable if, once bottled, it starts to ferment again, becomes cloudy, throws a deposit, or turns brown in less than 5 years of normal storage conditions.

Causes and Solutions

1. Insufficient sulphite. Always use the amount of sulphite recommended in our recipes. These are absolute minimum amounts and half what commercial wineries use.

2. Bottling before fermentation is complete. Always bulk store your wines for the period recommended in our recipes. Any sign of renewed fermentation will become evident during this time.

3. Malolactic bacteria contamination. Always bulk store for the recommended period. Any sign of contamination will become evident during this time.

4. Excessive oxygenation during bottling. Avoid splashing your wines during the bottling procedure. Make sure the syphon hose is at the bottom of the bottle.

5. Exposure to sunlight or fluorescent lights while maturing in clear glass bottles. Shield your wines from sunlight or fluorescent lighting. If you use clear glass bottles and you have no storage space, store your wines in cardboard tubes under a bed.

MYCODERMA (FLOWERS OF WINE)

The Problem

Mycoderma is a spoilage yeast that forms a film on the surface of wine — usually low-alcohol wine. It starts as islands of fine film that gradually thicken. When disturbed, these islands break up and drift down through the wine like snowflakes. Mycoderma quickly oxidizes the wine and produces a smell and taste that make it undrinkable.

The cause is most often too much airspace between the surface of the wine and the closure. Plastic screw caps are also culprits. Screw plastic caps on firmly, then take masking tape and run it around the closure to prevent the caps from loosening.

The Solution

If you catch mycoderma early enough (before the wine smells and tastes terrible), filter it with sterile-grade filter pads and add 50 ppm of sulphur dioxide (1 Campden tablet per gallon of wine). Having 30–50 ppm of sulphur dioxide in the wine at all times is the best prevention.

OXIDATION

The Problem

When an apple turns brown, it is oxidized. Oxidized wine is bitter and undrinkable. There is a theory that refermentation can reverse the oxidation process. But we suggest that if you have oxidized wine, pour it down the sink.

Strangely enough, oxidation is a problem for the knowledgeable winemaker as well as the beginner. The novice often does not realize the importance of topping up and sulphiting. The old hand gets careless and steals samples from barrels or carboys to taste the wine. He always tells himself he will top up later, but later is often a week later and the damage is done.

Wineries will move storage tanks of wine under a blanket of nitrogen or carbon dioxide to prevent oxidation, but this is not possible for winemakers who are making less than a thousand gallons. When we move small amounts, they are subject to oxidation if we do not take the proper precautions.

The problem of oxidation also crops up more often nowadays because of the increasing reluctance of home winemakers to use sulphite — despite the complete lack of evidence that it is hazardous when used in the manner and amounts advised.

Prevention

Use sulphite crystals every time you move a wine that is not fermenting from carboy to carboy.

Keep all wines topped up at all times.

STUCK FERMENT

The Problem in the Fermentor

At times a specific gravity reading will reveal that fermentation has failed to start.

The Solution

1. Check the temperature of the must. If necessary, adjust to 75°F (23°C) or slightly higher by moving it to a warmer location or by placing a heating pad under it. Stir it vigorously and cover.

2. If fermentation does not start within 24 hours, remove 1 cup of must and place in a glass bowl. Add 1 cup of warm water; sprinkle 1 packet (5 g) of yeast on the mixture and let it stand for 1 hour in a warm place. Then stir the mixture into the fermentor, cover and leave it in a warm location.

The Problem in the Carboy

All wines should be fermented out to dry — that is, all their sugar should be converted to alcohol — in order to attain a sufficient alcohol content for preservation. Wines can be sweetened to taste after fining. When a wine is fermented out to dry, its specific gravity will lie between .990 and .995.

When fermentation stops and there are no bubbles around the neck of the carboy, take a hydrometer reading. If the specific gravity is below .995, the process of fermentation is complete. If the specific gravity is above 1.020, proceed as follows:

The Solution

1. Check the temperature of the wine. If necessary, adjust it to 75°F (23°C) by moving it to a warmer location or placing a heating pad under it.

2. With the long handle of a wooden or plastic spoon, stir the yeast deposit into suspension and add 1 teaspoon of yeast nutrient. Reattach the fermentation lock to the carboy.

3. Take a second hydrometer reading after 3 days. If the specific gravity is reduced, allow the wine to continue fermenting at 75°F (23°C). If there is no change in the specific gravity, restart with fresh yeast.

Restarting with Fresh Yeast

The whole process of restarting with fresh yeast should not take longer than 1 week.

1. Assemble 6 clean gallon jugs, 6 fermentation locks, a syphon hose, and a 5-gram packet of dry wine yeast.

2. Place the dry wine yeast in a gallon jug (not sulphited) and add 1 cup of warm water (100°F [38°C]). Let stand 10 minutes.

3. Remove the cover or fermentation lock of the stuck wine and stir it well with the long handle of a wooden or plastic spoon.

4. Syphon the wine into the 5 remaining gallon jugs and attach fermentation locks to all but one. From the last jug, remove one cup of wine, then attach a fermentation lock.

5. Pour the cup of wine into the jug containing the yeast mixture. Reattach the fermentation lock and allow the mixture to ferment at a constant temperature of 75°F (23°C). Fermentation should take place within 24 hours.

6. When the yeast mixture is fermenting well, remove 2 more cups of stuck wine from the same jug as before and reattach its fermentation lock. Add the 2 cups of stuck wine to the fermenting yeast mixture and reattach its fermentation lock.

7. When the yeast mixture is fermenting well, add 2 more cups of stuck wine from the same jug. Replace the fermentation lock on the jug of stuck wine. Replace the fermentation lock on the yeast mixture.

8. Repeat the procedure, adding 2 cups of stuck wine as soon as the yeast mixture is fermenting well, until the jug of yeast mixture is full.

9. When the gallon jug of yeast mixture is fermenting well, remove the fermentation lock, stir the yeast into suspension, and syphon half of the yeast mixture into a clean (not sulphited) gallon jug.

10. You will now have 4 jugs of stuck wine and 2 half-full jugs of yeast mixture. Take one of the jugs of stuck wine and stir it well. Pour half of it into one jug of yeast mixture and half into the other. Attach a fermentation lock to each and allow the two jugs to ferment at a constant temperature of 75°F (23°C).

11. When you have 2 gallons of wine fermenting well, remove the locks on the remaining 3 jugs of stuck wine. Stir them well and syphon them into a clean (not sulphited) fermentor or carboy.

12. Remove the locks on the 2 fermenting jugs, stir the yeast into suspension, and syphon the fermenting mixture into the fermentor or carboy. Make sure you transfer all the yeast from the jugs into the fermentor or carboy. Cover the fermentor; attach a fermentation lock to the carboy.

13. Allow to ferment out at a constant temperature of 75°F (23°C).

Reductive Versus Oxidative Wines

When Dr. Elias Phiniotis took over our lab a few years ago, he brought some European expressions and techniques to our winemaking. One of the first things he did was separate our recipes into oxidative and reductive wines. Most people are familiar with the word *oxidize*; when things exposed to the air turn brown, such as iron or apples, they are oxidizing. The reverse of oxidation is reduction. Not all oxidation is bad and not all reduction is good. Properly controlled, both actions are needed in winemaking.

Reductive wines have a fresh fruity flavor and aroma, which has been deliberately retained. Most German wines are reductive, as are Portuguese Verde wines and many California wines — especially the whites and rosés. Reductive wines are permitted little or no oxygen and are usually consumed young — between 3 months and 3 years old.

Oxidative wines are aged in such a manner as to allow small amounts of oxygen into the wine over a long period of time — between 2 and 10 years. French reds, Spanish sherries, and Portuguese ports are oxidative. The oxidation process alters the flavor and changes the aroma to a bouquet.

The reductive/oxidative distinction is useful to winemakers when writing recipes and choosing such ingredients as acids, yeasts, and tannins. Reductive wines can be slightly lower in alcohol, higher in acid, and lower in tannin, and they usually benefit from a slight residual sugar. Also, they contain more malic acid than oxidative wines and are fermented at lower temperatures to preserve the flavor and aroma of the fresh fruit or grape.

Oxidative wines, such as the great Bordeaux and Burgundy reds, have different requirements: more alcohol, up to five times more tannin, higher fermenting temperatures, barrel aging, and if possible, long bottle aging in corked bottles. They also require different yeasts and acid additives. In oxidative wines, the aroma of the grape becomes only a small part of a much more complex bouquet. Better than reductive? No, but very different — more complicated and usually more expensive.

Refractometers

ALCOHOL REFRACTOMETER

The alcohol refractometer measures the alcohol content of a wine by its ability to refract light. We have checked readings from a refractometer against results obtained by distillation and an ebulliometer, and their accuracy is comparable. The refractometer is a great deal simpler to use, however, and can provide a reading within 90 seconds. In view of the fact that British government tax collectors consider refractometers accurate enough to assess alcohol tax, it is surprising that more wineries do not have them in their labs.

Equipment

A very accurate hydrometer (range: .0975–1.025)

Glass hydrometer jar, 15 inches (38 cm) tall

Floating thermometer

Eye dropper or glass rod

Bellingham and Stanley light wine refractometer (range 9–25% alcohol by volume)

Procedure

1. Open the hinged prism box of the refractometer and place a couple of drops of wine on the lower prism. As you close the prism box, a film of wine is created between the two prisms.

2. Direct the instrument toward a light source, look through the eyepiece, and adjust the focus until you get a sharp image. Your field of view will consist of a numbered scale divided into dark and light. You simply read off the point on the scale where dark and light meet. This is the refractive index of your wine.

3. Take an accurate hydrometer reading of the wine. If the specific gravity is less than 1.000, you add the number by which it is less than 1.000 to the refractive index. If the specific gravity is more than 1.000, you subtract the number by which it is more than 1.000 from the refractive index.

4. Look up the resultant figure in the tables supplied with the instrument and you have the alcohol content of the wine.

Two precautions must be observed: The instrument must be kept perfectly clean and the temperature of the wine sample must correspond to the temperature calibration of the hydrometer. The refractometer can be adjusted for temperature, but the hydrometer cannot.

An alcohol refractometer costs about $200, but wine clubs or small groups of winemakers should find one worthwhile. For information about sales outlets, see "Buyer's Guide" (page 281).

SUGAR REFRACTOMETER

The sugar refractometer operates on the same principle as the alcohol refractometer: It measures the concentration of a sugar solution by its ability to refract light. Thus, it does the same job as a hydrometer, but because it uses only two drops of grape juice, it can be used by the vineyardist in the field to decide when to harvest. An alcohol refractometer costs about $100.

Residual Sugar

After fermentation has stopped and the specific gravity reading is .995 or less, you may assume that there is no reducing (fermentable) sugar left in the wine. If the original SG was above 1.100, you could be wrong and proceed to bottle an unstable wine with residual sugar. This would ferment in the bottles and blow the corks. There are two commercial products available to check for residual sugar. One, Clinitest, is designed for diabetics; the other, Dextrocheck, is designed for winemakers. Both indicate the concentration of sugar by progressive change, from blue (no sugar) to bright orange (2% residual sugar).

In a completely dry wine, a residual sugar reading of ½% usually indicates nonfermentable polysaccharides that can be considered stable and safe for bottling. But a wine with a reading of more than ½% should be fine-filtered and stabilized with 30 ppm sulphur dioxide and 200 ppm potassium sorbate. If the wine has more than 2% residual sugar, and you find it too sweet, you should consider refermenting it. (See "Stuck Ferment," page 269.)

Sanitizing Solutions

Two chemicals are used in the wine industry to clean equipment and destroy undesirable fungus or bacteria. They are potassium metabisulphite and quartenary bactericide. The solutions are made up in the following formulas.

Sulphite solution: 1¾ ounces or 4 tablespoons (50 g) of potassium metabisulphite dissolved in 1 gallon (4 lit) of warm water

Bactericide solution: 6 teaspoons (30 ml) quarternary bactericide mixed into 1 gallon (4 lit) of water

Fifteen years ago we called these products sterilizing agents, but the U.S. government ruled that the word *sterile* or *sterilizing* implied a solution strong enough to be a health hazard, even though these solutions kill only undesirable fungus and bacteria. They are now referred to as sanitizing solutions.

The bactericide is stronger and more effective than the sulphite solution, but it must be rinsed out of containers completely with fresh water. It cannot be used in fermentation locks, and if any residue of the bactericide comes in contact with the wine, it will spoil the wine.

The sulphite solution, on the other hand, is nontoxic, and any trace left in containers is actually beneficial to the wine. Sulphite solution is also used in fermentation locks. After using either sanitizing solution to rinse out carboys and clean equipment, you can pour back any excess into your gallon storage container and use it again. Every time sanitizing solution is used, however, it loses some of its strength; therefore, make a practice of discarding a solution after you have used it 10 times. Keep your storage container tightly capped at all times.

Substitutes

In North America, winemaking products are widely available; but if you are in a situation of dire emergency, we offer these substitutes.

Ingredient	**Substitute**
SUGAR (white, or cane, is best for winemaking)	Honey — use 30% more. Corn sugar — use 20% more.
GRAPE TANNIN	Tannic acid.
VINACID R and O	Acid blend — use an equal amount. Lemon juice — use twice as much.* Citric acid — use 20% less.
GRAPE CONCENTRATE	Raisins (they will give a sherry taste to the wine) — use 8 oz raisins to 10 oz concentrate.
FERMENTATION LOCK	Plastic bag — place over the neck of the fermentor and secure with a rubber band. Rubber balloon.
PRIMARY FERMENTOR	Plastic garbage tub. Garbage-tub plastic could be toxic, so move to a food-grade plastic tub as soon as possible.
SECONDARY FERMENTORS	Cubitainers (tend to oxidize wine).
STRAINING BAG	Nylon stocking.

*Will vary with lemons used.

Exact Measurements

Vinacid: 1 level teaspoon = 4 g = 5 ml

Sulphite crystals: 1 level teaspoon = 6 g = 5 ml

Tannin: 1 teaspoon of liquid tannin = 5 ml (For 1 teaspoon of liquid tannin, use 2 teaspoons of dried tannin.)

Syrup Feeding

By syrup feeding, you can make high-alcohol aperitif, dessert, or flavored wines without resorting to fortification with expensive spirits. We have tested cherry wine with 19.8% alcohol produced by this method. The general principles of syrup feeding are as follows:

1. Select an alcohol-tolerant yeast.
2. Adjust the starting specific gravity of your must to 1.090.
3. Cool-ferment your wine at 60–70°F (15–21°C) to SG 1.000.
4. Prepare a sugar syrup by adding sugar to boiling water in the proportions of 2 cups of sugar to 1 cup of water. When the sugar is dissolved, allow the solution to cool to room temperature.
5. When your wine has been racked into carboys, add the syrup until the specific gravity rises to 1.010. Allow it to continue cool-fermenting at 60–70°F (15–21°C). Yeast will tolerate a higher alcohol concentration at cooler temperatures.
6. When the specific gravity drops to 1.000, bring it up to 1.010 with syrup. You can repeat this 4–5 times. One pound of sugar added to 5 gallons of must will raise the specific gravity .010.
7. If you are anxious to avoid too sweet a wine, or are concerned about a stuck ferment, bring the SG back only to 1.005 after your second or third feeding. However, you should note that aperitif wines with a high alcohol content need some residual sweetness. For example, a sherry with an alcohol content of 17–18% by volume and an SG of 1.000–1.005 will taste quite dry, even though it may have 3–4% residual sugar.

Tasting and Evaluating Wine

The Davis campus of the University of California developed a system of evaluating wine that makes it possible for almost everyone who is interested to make a serious assessment of any wine. It removes most of the snobbery and esoteric language of the European and English masters of wine and lends itself to a more literate description and practical analysis. We do not wish to deprecate the great masters of wine, but unfortunately it would take most of us a lifetime to learn the language of their profession, not to mention the thousands of bottles of wine that must be sampled.

Many organoleptic (judging) evaluations are based on numerical value, usually 1–100. But Davis discovered that most judges use only 10 points to differentiate between good and bad wines; so they devised a system using only 20 points.

The Davis system was developed more than 20 years ago, so the numerical value assigned to each attribute has been altered slightly to reflect the improved technology of modern winemaking. For example, the original sheet gave 2 points for clarity; yet we cannot remember the last time we bought a cloudy wine in North America, so we do not think it is an important point to assess. Frankly, if anyone served us a cloudy wine today, we would refuse to drink it.

It's not only useful to define exactly the kind of wine you like, but satisfying to be able to explain why you personally think it is a good sound wine.

DEFINITION OF TERMS

Appearance, or Clarity: 1 Point

A slight firm sediment is not objectionable; but if it prevents you from decanting clear wine, it merits the loss of a point. Thus, a perfectly clear wine (brilliant or star bright) rates 1 point; a slightly opalescent wine, or a wine with a loose sediment, rates ½ point; a cloudy wine, or a wine with suspended particles suggesting contamination, rates 0 points.

Aroma and Bouquet: 5 Points

Aroma is the smell of the ingredients; bouquet is what develops during aging if the wine is properly made. New wines seldom have bouquet; therefore, if you are judging a new wine, you would not fault it for lack of bouquet if you thought it had potential. Thus, a varietal wine from Muscat grapes with a distinct aroma of the grape would merit 3; if it also had a vinous bouquet, it could rate 5. A wine lacking a pronounced or identifiable aroma or bouquet, but without any unpleasant smell, rates 2. A wine with a slight off odor, such as a hint of hydrogen sulphide or a rubbery smell, rates 1. A wine that has vinegary or spoiled odors rates 0.

Astringency, or Tannin: 1 Point

Tannin gives wine character. Too much makes a wine harsh and rough; too little makes a wine flabby and insipid. Astringency is easily confused with high acidity and is often masked by high alcohol and residual sugar. It is more pronounced and acceptable in red wine than white, and young wines often have an excess. In young wines, high tannin is not necessarily a fault, since it may disappear as the wine ages.

Body: 2 Points

Body is a feeling of fullness on the palate that is not caused by sugar. It is particularly important in red table wines; white table wine, on the other hand, can be fairly thin. A wine with a body in balance with its type merits 2 points; body that is medium but adequate rates 1 point; a weak-tasting wine rates 0 points.

Color: 1 Point

Example: A white wine can vary from white to straw gold and rate 1; if it is a deep golden brown, it rates ½ point; if it is dark brown or reddish, it rates 0.

Flavor and Balance: 5 Points

Characteristics that lose points are bitterness, yeastiness, metallic flavor, caramelization, and an overly alcoholic taste (hot, dry sensations). Terms such as *young, tired,* or *bottle ripe* are also used to describe flavors. Do not confuse this category with the hedonist rating that concludes the list. You do not have to like vermouth to be able to recognize a good one.

Sugar: 1 Point

In red table wines you would not expect to detect sweetness; if you do, award no points. In white table wines not specifically designated dry, 1–2% sugar is acceptable and would merit 1 point. In cream sherry, you would expect 8–12% residual sugar and a definite sweetness; here a lack of sugar would cost a point.

Total Acid: 2 Points

What is considered a desirable total acid level depends on the type of wine. For example, sherry with 4–5 parts per thousand would rate 2. A wine slightly high or low would rate 1. If a wine is very high — say, a red with 10 parts per thousand — award it 0.

General Quality: 2 Points

In this category you can express your personal reaction to the wine. If you wish you had made it yourself, or wish you know where to buy it, give it 2 points. If it has a definite defect, and you would not want to drink or serve it, give it 0 points.

Wine Appraisal Sheet

	Possible points	*General comments*	*Points given*
Clarity	1		
Aroma and bouquet	5		
Astringency (tannin)	1		
Body	2		
Color	1		
Flavor and balance	5		
Total acid	2		
Sugar	1		
General quality	2		

Total possible points 20 Total points awarded ______________

Name of wine ______________________ Date bottled ______________

WINE COLORS AND DESCRIPTIVE WORDS

There are three color categories of wine — red, rosé, and white — but there is a wide spectrum of color within each category. Red wines can be deep purple, ruby red, red, red brown, deep brown, or amber. Blue tints in red wine suggest youth; brown tints in red wine suggest age, and if the wine is sound and well aged, they are not a fault. Often, however, brown table wines are oxidized and have a bitter taste and cooked flavor.

Rosé wines range in color from onion skin to pink to orange. White wines can be palc green, straw yellow, gold, yellow brown, or brown. Sherry is in the white-wine category.

The clarity of a wine is described as star bright, brilliant, clear, cloudy, dull, or watery.

Following are some of the words used to describe wines:

Bouquet and Aroma

Acetic	Acidic	Beery
Blackcurrant	Clean	Corky
Dominant	Flinty	Flowery
Foxy	Fruity	Geranium
Green	Honey	Madeira
Musty	Peppery	Powerful
Smoky	Spicy	Sulphury
Woody	Yeasty	Rubbery

Taste

Astringent	Sharp	Green
Mellow	Flat	Thin
Bitter	Sound	Insipid
Peppery	Fruity	Yeasty
Corky	Sweet	Madeira
Balanced	Tart	Smooth
Oxidized	Honey	Flinty
Coarse	Woody	Spicy
Rounded	Baked	Full
Earthy	Metallic	Tannic
Soft	Clean	Harsh
Foxy	Ripe	Velvety
Sulphury	Dry	Light
Grapey		

GLOSSARY

ACID BLEND Mixed fruit acids to add to deficient musts

ACID CRYSTALS Potassium bitartrate deposit — often called wine stone or cream of tartar

AGING Storing finished wine so that it can mature

ALKALI Chemical opposite of acid; e.g., chalk

AMELIORATION Balancing a must by adding sugar, water, or grape concentrate

ANAEROBIC Without air

ANTIOXIDANT Agent that prevents oxidation spoilage; e.g., sulphur dioxide, sodium erythorbate

ASTRINGENCY Effect of tannin; puckery sensation

AUTOLYSIS Decomposition of yeast cell proteins to amino acids as a result of action by the cell's own enzymes

BAKE Long-term exposure of wine to an elevated temperature in order to make Madeira or American sherry

BALLING Hydrometer scale used in America to measure sugar content

BANANA POWDER Additive used to increase body in wine recipes

BATCH Quantity of wine produced at one time

BODY The characteristic of wine that gives a sensation of fullness to the palate

CAMPDEN TABLET ½-gram tablet of metabisulphite

CAP Fruit skins and pulp floating on top of fermenting wine

CARBON DIOXIDE Gas released by fermentation, present in champagne

CHILLPROOFING Removing acid from wine by chilling

CITRIC ACID Acid found in lemons and other citrus fruit; sometimes added to wine or must

CORDIAL Sweet fruit beverage, 20–24% alcohol

DEXTROSE Monosaccharide (sugar) from corn — not very sweet

ELDERBERRIES Additive used to give color and tannin to wine

ELDERFLOWERS Additive used to give aroma to wine

ENOLOGY Science of wine

EXTRACT Nonvolatile flavor base for wine or liqueur

FALSE WINE Wine made from the second or third fermentation of a single quantity of grapes

FERMENTATION Conversion of sugar into alcohol and carbon dioxide by yeast, noticeable as bubbles rising in must

FININGS Clearing agent, such as gelatin, bentonite, etc.

FIRST RUN First fermentation of grapes (no water or sugar added)

FLOR Surface-growing yeast used to ferment sherry

FREE RUN Juice released from grapes by crushing

FULL-BODIED Good, satisfying; characteristic of a big wine with a lasting mouth sensation

HYBRID GRAPES A cross between North American and European grapes

HYDROGEN SULPHIDE Rotten-egg smell that can develop in wine from yeast autolysis

GLYCERINE Agent used to sweeten and add body to wine

GYPSUM Agent used in sherry to lower pH of must

LACTIC ACID Result of malolactic fermentation

LEES Sediment at bottom of fermentation vessel

MALIC ACID Acid found in apples, grapes, etc.

MALT EXTRACT Agent used to provide body and nutrient in some sherry recipes

METABISULPHITE Antioxidant and bactericide that releases sulphur dioxide into wine

MINERAL Iron or copper casse causing cloudy wine

MUST Crushed fruit or juice for fermentation

MYCODERMA Flowers of wine, a spoilage organism

OAK CHIPS, STICKS Oak fragments used to add oak flavor to wine

OXIDATION Process by which wine is spoiled by exposure to air

OXIDATIVE WINES Wines that need to absorb small quantities of oxygen during a lengthy aging

PECTINASE Pectic enzyme to remove the pectin cloud from wine

pH Scale of acidity and alkalinity

PHENOLPHTHALEIN Agent that reveals the presence of acid by a change in color

POMACE Pressed grape solids (skins)

POTASSIUM SORBATE Stabilizing agent that prevents renewed fermentation in sweetened wine

POTENTIAL ALCOHOL Amount of alcohol by volume expected after fermentation is complete

PRESSING Squeezing juice from crushed fruit

PRIMARY FERMENTATION First 5–6 days of vigorous fermentation

PULP Crushed fruit or grape solids

QUICK WINES Recipes that produce wine in 28 days or less

RACKING Syphoning wine from one container into another

REDUCTIVE WINES Wines that are permitted little or no oxygen and are consumed young

REHYDROLYSATION The activation of yeast in warm water

RESIDUAL SUGAR Sugar remaining in wine after fermentation has stopped

SANITIZERS Bureaucratic euphemism for sterilizer; e.g., sulphite solution

SCRUMPY Old-fashioned apple cider or perry

SECONDARY FERMENTATION Fermentation in a carboy or barrel from which air is excluded

SECOND WINE Wine made in a second run

SINATIN 17 Oak essence to provide flavor and hasten wine aging

SODIUM ERYTHORBATE Isomer of vitamin C used as an antioxidant

SODIUM HYDROXIDE Alkali used in acid titration

SOUND WINE Wine with no obvious defect

SPECIFIC GRAVITY (SG) Scale of density used to measure sugar in must or wine

SPLASHING Aerating wine to release carbon dioxide gas

STABILIZER Potassium sorbate, used to prevent renewed fermentation in sweetened wine

STABLE Term used to describe a clear wine in which all fermentation is complete

STARTER High concentration of yeast cells to inoculate wine

SUCCINIC ACID An acid created in wine during fermentation

SUCROSE White granulated cane or beet sugar

SULPHITE Sodium or potassium metabisulphite

SULPHUR DIOXIDE Antioxidant gas released into wine by metabisulphite

SWEET WINE Wine with more than 2% residual sugar

TABLE WINE Wine intended to be consumed with food; usually dry

TANNIN Astringent-tasting phenol from grape seeds and skins

TITRATABLE ACID Acid measured by titration as grams per liter

TOPPING UP Adding wine or water to fill the secondary fermentor

VINACID O Trade name for blended acids suitable for oxidative wines, sherries, and ports

VINACID R Trade name for blended acids suited to reductive white wines, rosés, and fruit wines

VINIFERA The true wine grape; there are thousands of varieties

VINOUS Winelike; tasting or smelling like good wine

VISCOSITY Stickiness

VOLATILE ACID Acid created during fermentation or spoilage; for example, vinegar

WINE CONDITIONER Mixture of liquid invert sugar and sorbic acid, used to sweeten finished wine

YEAST NUTRIENT A mixture of vitamins and minerals added to a must to ensure healthy fermentation

BUYER'S GUIDE

CATALOGUES — RETAIL OUTLETS

Canada

BREW KING*
2540D Shaughnessy
Port Coquitlam, B.C. V3C 3G1

BREWMASTER SYSTEMS, Ltd.*
106–250 Eighteenth Street
West Vancouver, B.C. V7V 3V5

BREW-IT-YOURSELF
261-C Trans Canada Highway
Duncan, B.C. V9L 3R1

CASK BREWING SYSTEMS
3672 Sixtieth Avenue SE
Calgary, Alberta T2C 2C7

CASK BREWING SYSTEMS
35 King Street North
Waterloo, Ontario N2J 2W9

*Head office — write for address of store nearest you

SPAGNOL'S Wine and Beermaking Supplies, Ltd.
1325 Derwent Way
New Westminster, B.C. V3M 5V9

WINE ART, Inc.*
250 West Beaver Creek Road
Units 8–10
Richmond Hill, Ontario L4B 1C7

WINE ART, Inc.*
6080 Airport Road South, Unit 120
Richmond, B.C. V7B 1B4

United States

ANDREWS HOMEBREWING ACCESSORIES
5740 Via Sotelo
Riverside, CA 92506

BACCHUS & BARLEYCORN, Ltd.
8725 Johnson Drive
Mission, KS 66202

BEER AND WINE HOBBY
PO Box G104 Greenwood
Wakefield, MA 01880

BREWER'S & WINEMAKER'S MARKET
Box 298Z
Sumner, IA 50674

CAPE COD BREWERS SUPPLY
126 Middle Road, Box 1139C
South Chatham, MA 02659

CROSBY & BAKER
999 Main Road
PO Box 3409
Westport, MA 02790

DOVER VINEYARDS
24945 Detroit Road
Cleveland, OH 44145

F. H. STEINBART Co.
602 SE Salmon
Portland, OR 97214

GREAT FERMENTATIONS
87 Z Larkspur
San Rafael, CA 94901

HOME-BREW INTERNATIONAL
Box 4547-4Z
Fort Lauderdale, FL 33338

JOE and SONS ZY
PO Box 11276
Cincinnati, OH 45211

SEMPLEX
Box 11276
Minneapolis, MN 55411

THE BRASS CORKSCREW, Inc.
PO Box 30933
Seattle, WA 98103-0933

THE CELLAR
Dept ZD, PO Box 33525
Seattle, WA 98133

THE WINE AND HOP SHOP
705 E Sixth Avenue
Denver, CO 80203

WINE ART INDIANA
5890 N Keystone Avenue
Indianapolis, IN 46220

WINEMAKING SHOPPE
Route 1, Box 64W
Sugar Grove, IL 60554

WINES, Inc.*
1340 Home Avenue
Akron, OH 44310

KOEPPL'S MASTER BREWER
2311 George
Rolling Meadows, IL 60008

KRAUS
Box 7850-Z
Independence, MO 64053

NORTH DENVER CELLAR
4370 Tennyson
Denver, CO 80212

*Wines, Inc. has many outlets across the U.S. Call 1-800-321-0315 for the retail store nearest you.

OREGON SPECIALTY COMPANY, Inc.
7024 NE Glisan Street
Portland, OR 97213

PERTZ
PO Box 478
Endicott, NY 13760

THE PURPLE FOOT
3167 South Ninety-Second Street, Dept Z
Milwaukee, WI 53227

ROOT & VINE
Box C51Z
Westport, MA 02790

BOOKS AND MAGAZINES

THE WINE APPRECIATION GUILD
155 Connecticut Street
San Francisco, CA 94107
Catalogue available $1.00
Over two hundred titles

WINES & VINES
1800 Lincoln Avenue
San Rafael, CA 94901

MAGAZINES ONLY

BEVERAGE COMMUNICATOR
Box 43
Hartsdale, NY 10530

TIDINGS
5-2140 Grey Avenue
Montreal, Quebec H4A 3N4

VINTAGE MAGAZINE
PO Box 866
Madison Square Station
New York, NY 10010

WINE EAST
620 North Pine Street
Lancaster, PA 17603

ZYMURGY
PO Box 287A
Boulder, CO 80306

EQUIPMENT

Alcohol Refractometer

Bellingham & Stanley, Ltd.
61 Markfield Road
London N15 4QD
England

Sulfikit

Hazlemere Research
1940 180th Street RR8
Surrey, B.C. V3S 5J9
Canada

FRESH GRAPES AND FRESH JUICE*

Canada

SPAGNOL'S
1325 Derwent Way
New Westminster, B.C. V3M 5V9

*Information is also available from Department of Agriculture in the wine-growing areas of Canada and USA.

WINE ART, Inc.
Information for across Canada
Head Office
250 West Beaver Creek Road
Units 8–10
Richmond Hill, Ontario L4B 1C7

United States

AMERICAN WINE SOCIETY
3006 Latta Road
Rochester, NY 14612

OREGON WINEGROWERS ASSOCIATION
PO Box 6590
Portland, OR 97228-6590

WINES, Inc.
1340 Home Avenue
Akron, OH 44310

GRAPE CUTTINGS AND PLANTS*

Canada

GRAPE GROWERS ASSOCIATION
12-436 Bernard Avenue
Kelowna, B.C. V1Y 6N8

United States

CALIFORNIA NORTH COAST VINIFERA WINE GRAPES
Ship out-of-state orders
(707) 433-1944 or (707) 545-2310

FOSTER NURSERY CO., Inc.
69 Orchard Street
Fredonia, NY 14063

SONOMA GRAPEVINES, Inc.
1919 Dennis Lane
Santa Rosa, CA 95401

SQUARE ROOT NURSERY
4764 Deuel Road, Dept W
Canandaigua, NY 14424

OTHER SUPPLY SOURCES

For the names, addresses, and phone numbers of other suppliers of equipment and ingredients in the U.S. and Canada, contact:

HOME WINE & BEER TRADE ASSOCIATION
604 North Miller Road
Valrico, FL 33594
813-685-4261
Fax 813-681-5625

*Information is also available from Department of Agriculture in the wine-growing areas of Canada and USA.